Economic Geology and Geotectonics

Economic Geology and Geotectonics

Edited by D. H. Tarling
BA, PhD, FGS, Reader in Palaeomagnetism,
Department of Geophysics and
Planetary Physics,
The University, Newcastle-upon-Tyne

A Halsted Press Book

John Wiley & Sons
New York Toronto

© 1981 by Blackwell Scientific Publications

All rights reserved. No part of this
publication may be reproduced, stored
in a retrieval system, or transmitted,
in any form or by any means,
electronic, mechanical, photocopying,
recording or otherwise
without the prior permission of
the copyright owner

First published in 1981

Published in the U.S.A. and Canada
by Halsted Press
a Division of John Wiley & Sons, Inc.,
New York

Library of Congress
Cataloging in Publication Data

Economic geology and geotectonics.
  1. Geology, Economic   2. Geology,
Structural
I. Tarling, Donald Harvey
553   TN260

ISBN 0 470 27145 0

Printed in Great Britain

Contents

List of Contributors, vii

Preface, ix

Chapter 1 Introduction: Plate Tectonics *by D. H. Tarling*, 1
- 1.1 The basic theory, 1
- 1.2 Constructive (accretionary) plate margins and the oceanic lithosphere, 4
- 1.3 Destructive plate boundaries (subduction zones) and the continental lithosphere, 9
- 1.4 Continental separation during the last 300 million years, 13
- 1.5 The mechanism of plate tectonics, 16
- 1.6 Past plate tectonics, 20

Chapter 2 Petroleum: Introduction and the Formation and Migration of Hydrocarbons *by R. Stoneley and R. J. Bailey*, 31
- 2.1 General introduction, 31
- 2.2 Environmental factors in the deposition of petroleum source facies, 34
- 2.3 The formation of petroleum, 42
- 2.4 The migration of petroleum, 44

Chapter 3 Petroleum: the Sedimentary Basin *by R. Stoneley*, 51
- 3.1 The classification of basins, 51
- 3.2 Basins associated with crustal spreading, 52
- 3.3 Basins associated with crustal destruction, 57
- 3.4 Basins associated with transform faulting, 61
- 3.5 Intraplate basins, 64
- 3.6 Conclusions, 70

Chapter 4 Petroleum: Entrapment and Conclusions *by R. J. Bailey and R. Stoneley*, 73
- 4.1 Reservoirs, 73
- 4.2 Petroleum traps, 80
- 4.3 Cap-rock or seal, 90
- 4.4 The deep-water environment, 91
- 4.5 Petroleum: general conclusions, 94

Chapter 5 Coal *by G. A. L. Johnson*, 99
- 5.1 Introduction, 99
- 5.2 The process of coalification, 101
- 5.3 The depositional environment of coal, 104
- 5.4 Coal formation and plate tectonics, 114

Contents

Chapter 6   Ore Deposits of the Ocean Crust *by J. R. Cann*, 119

       6.1 Introduction, 119
       6.2 Oceanic crustal processes, 119
       6.3 Magmagenic ore deposits, 123
       6.4 Hydrothermal ores, 127
       6.5 Manganese nodules, 132

Chapter 7   Ore Deposits Associated with Subduction *by R. D. Beckinsale and A. H. G. Mitchell*, 135

       7.1 Introduction and historical background, 135
       7.2 Magmatism and mineralisation at convergent plate margins related to tectonic settings, 135
       7.3 Mineralisation in magmatic arcs, 137
       7.4 Mineralisation in back-arc settings, 140
       7.5 Mineralisation in outer-arc settings, 143
       7.6 Mineralisation in collision belts, 143
       7.7 Tectonic settings in relation to mineral exploration, 144

Chapter 8   The Origins of Ore Deposits in Sedimentary Rocks *by J. P. N. Badham*, 149

       8.1 Introduction, 149
       8.2 Primary sedimentary deposits, 151
       8.3 Secondary sedimentary deposits, 159
       8.4 Conclusions, 184

Chapter 9   Palaeoclimatic Considerations and General Conclusions *by D. H. Tarling*, 193

       9.1 Introduction, 193
       9.2 Palaeoclimatic factors, 193
       9.3 General conclusions, 201

Index, 207

List of Contributors

Dr N. Badham, *Department of Geology, University of Southampton*

Dr R. J. Bailey, *BP, Britannic House, Moor Lane, London EC2*

Dr R. D. Beckinsale, *Geochemical Division, IGS, 64 Gray's Inn Road, London WC1*

Professor J. R. Cann, *Department of Geology, The University, Newcastle-upon-Tyne*

Dr G. A. L. Johnson, *Department of Geology, University of Durham*

Dr A. H. G. Mitchell, *Mineral Exploration Project, PO Box 107, Lainchour, Kathmandu, Nepal*

Professor R. Stoneley, *Department of Geology, Imperial College, London SW7*

Dr D. H. Tarling, *Department of Geophysics and Planetary Physics, The University, Newcastle-upon-Tyne*

Preface

It is difficult to specify a single ore body, coal or oilfield that has, so far, been discovered by means of the direct application of the new plate tectonic theory. The revolution in the Earth Sciences which this theory has caused since the mid 1960s is not, therefore, of immediately obvious relevance to exploration geology. One reason for this is that it is always difficult to determine when a change in conceptual outlook actually becomes pragmatically significant. So far the main impact of the theory has probably been to provide a causative relationship between the previously disparate academic branches of the Earth Sciences—geophysics, geochemistry, petrology, palaeontology, stratigraphy, etc.,—but an obvious common interest already unified these subjects within an industrial context. Any delay in its impact can be largely attributed to other factors, particularly politico-economic, often being of greater immediate consideration than purely geological factors. In any case there is inevitably a long delay between the initiation of any exploration programme and its fruition. In the long term, it seems probable that the main contribution will be establishing a more realistic framework within which the origin of economically important concentrations can be better evaluated, but it is already possible to evaluate certain aspects of the new global tectonic models that are immediately relevant to Man's future search for geological resources. Such an assessment seems particularly opportune when there is an increasing awareness of the finiteness of such reserves and hence of the need for increasingly sophisticated, and hence increasingly expensive, exploration techniques. In such a context, models of present and past global tectonic processes are likely to lead to a better definition of areas in which such techniques are likely to be most successful at the same time as providing a better understanding of their mode of formation and hence of the optimum methods for both their detection and extraction.

It is thought that the assessments within this book will be of direct value in the planning and evaluation of exploration programmes, as well as being directly relevant to the interests of final-year and postgraduate students. As the level to which global tectonics and economic geology is taught varies considerably, only a limited knowledge of these subjects

Preface

could be assumed and it is hoped, therefore, that this book will also provide a general introduction for Economic Geology and thus be complementary to standard texts that provide detailed, local background to individual economic concentrations. Such a book cannot, of course, be comprehensive and there is no discussion of major resources such as water, gravel, sands, phosphates, etc. The relevance of global tectonics to some of these resources is, in any case, small—a knowledge of the cause of the last Ice Age is unlikely to lead to the discovery of major new sources of sand and gravel in areas where such materials are scarce already. However, as in the case of phosphates, gypsum, diamonds, asbestos and most other industrial minerals it is thought that the principles and ideas outlined here can also be readily applied to other types of ore.

The basic concepts of plate tectonics, as operating during the last 200-300 million years, are outlined in Chapter 1, together with a brief discussion of the possible mechanism and hence an assessment of the form in which such processes may, or may not, have operated during more remote times. The origin of hydrocarbons, their release from their source rocks, migration and entrapment are considered by Professor R. L. Stoneley and Dr R. J. Bailey. The formation of peat and eventually coal is outlined by Dr G. A. L. Johnson with particular reference to the formation of the Laurasian and Gondwanan coals. The formation of metal deposits along the oceanic ridges and within ocean basins is discussed by Professor J. R. Cann, and those in subduction zones by Dr R. D. Beckinsale and Dr A. H. G. Mitchell. Although such metallic ores are primarily associated with igneous activity, some major ore bodies are now associated with sedimentary environments and these types are considered by Dr N. Badham.

There are overlaps between sections of the chapters of some authors and these have largely been retained on two grounds. First, that a division of ores in terms of the tectonic context within which they are now found corresponds closely, but not exactly, with a division in terms of their mode of formation (if known). The second reason is that there are clear distinctions between the views of the different authors and it is felt healthier that such differences should be evident rather than to give an impression that the advent of new global tectonic concepts has solved the problems of the origin of hydrocarbons, coal or metallic ores when they have merely provided a more realistic framework within which both new and old ideas can be better evaluated. Such an evaluation is attempted in the concluding chapter, but is hoped that the reader will, in fact, make their own assessment.

D.H.T.

Chapter 1

Introduction: Plate Tectonics

D.H. TARLING

1.1 The basic theory

The basic concept of plate tectonic theory is elegantly simple. It proposes that new oceanic crust is being created along a narrow zone at the crest of the world's seismically active oceanic ridges (Fig. 1.1). These new oceanic crustal rocks move perpendicularly away from the ridges together with an increasing thickness of the Earth's upper surface which may, or may not, include continental crustal rocks. The creation of new oceanic crust is compensated by its destruction as it is returned into the mantle along the lines of the world's oceanic trenches (Fig. 1.1).

The areas of the Earth's surface that are moving as coherent units are termed 'plates' as they are considered to have much greater lateral extent than vertical thickness. Each plate is tectonically quiet away from its boundaries because it is only along the margins of plates that there is major differential motion, and hence seismic and tectonic activity (Le Pichon, 1968;

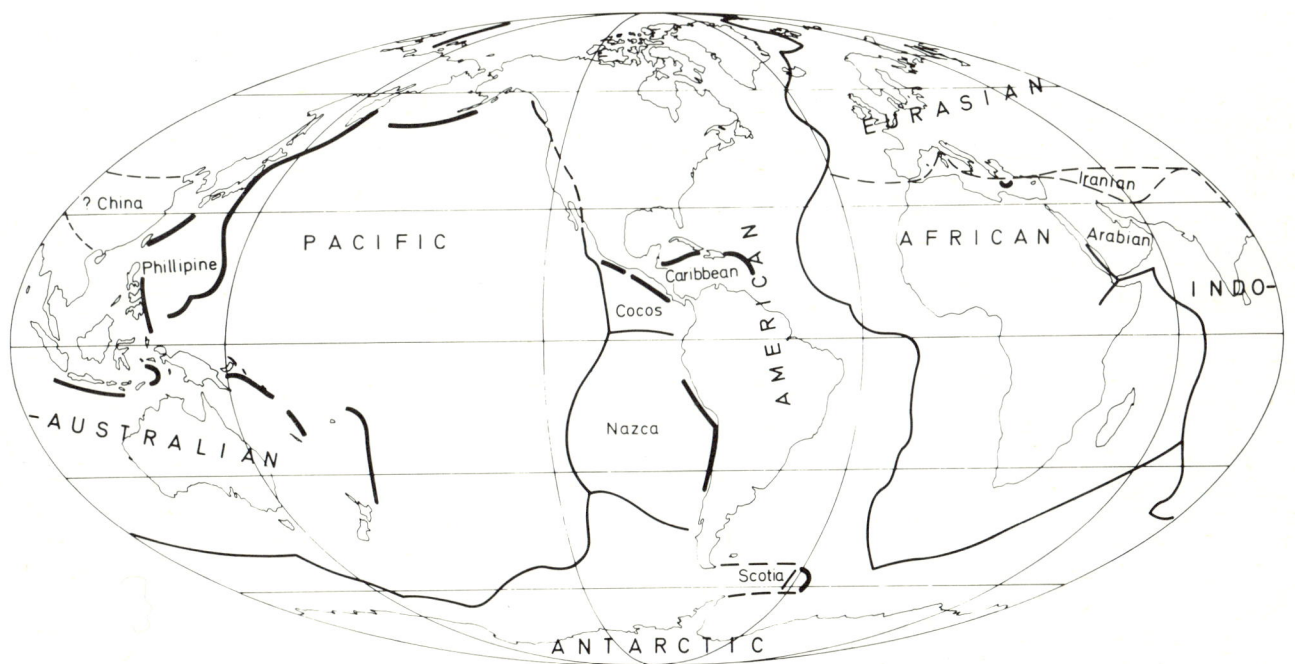

Fig. 1.1. Tectonic plates of the world today. Major earthquakes define plate edges as 80% of these are generated by the differential motion between the 'rigid' plates of the Earth's surface. Eight major plates can be readily identified, but numerous small plates can be recognised as having individual motions. Some of these may only reflect reactions of small tectonic units caught between two or more major plates, while others may have their own driving mechanism which results in their independent motion (after Tarling, 1978).

Chapter 1

Morgan, 1971). This activity is radically different at constructive (accretionary) plate margins, i.e. where new oceanic crust is being created, to that at the destructive plate margins where oceanic rocks are subducted. At the constructive margins the seismic activity is tightly confined and restricted to narrow, shallow zones at the crest of the ridges and the tectonics are essentially of tensional rifting (Fig. 1.2). At the subduction zones, the seismic energy released is much greater and the active zone dips into the Earth down to maximum depth of 600-700 km (Fig. 1.3). Overlying the dipping earthquake zone at such destructive margins, there is intense volcanic and tectonic activity. This may be associated with an active island arc system, for example the Western Pacific, or with strong orogenic activity along a continental edge, for example the Andes. Plate margins may also occur as narrow transform fault zones in which tectonic and igneous activity is low and the seismic energy release, while greater than that along constructive boundaries, is small and confined to narrow, shallow zones.

A plate tectonic model for the Earth's surface today incorporates some eight major plates, with numerous smaller plates. Some microplates appear to have an independent motion while others appear to have motions that are controlled by the movement of the larger plates that they border. The former are mostly of oceanic rocks and the latter are of continental rocks, usually in orogenic belts. Clearly such a model must evolve and must also be capable of predicting past and future distributions of both oceanic and continental plates. While there seems to be no clear understanding why constructive margins occur where they do, the location

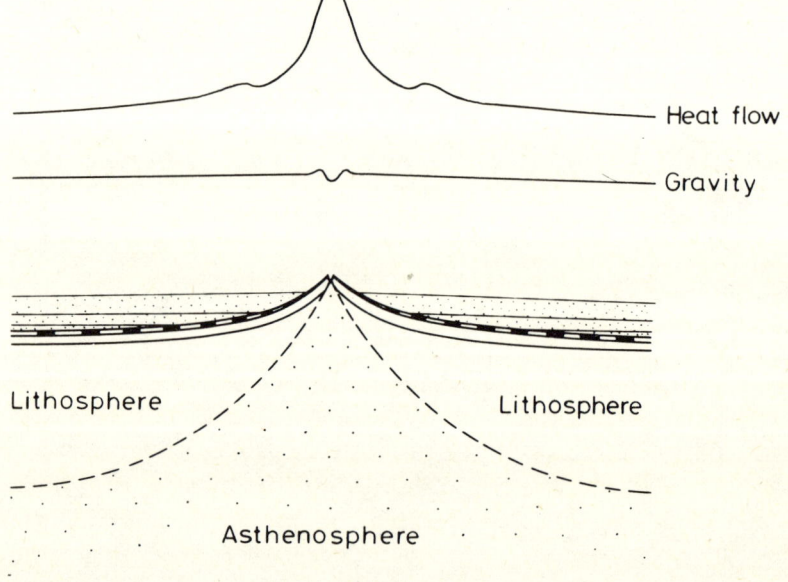

Fig. 1.2. A model of an oceanic spreading ridge. New oceanic crust is generated at the crest of the ridge where it cools and is magnetised in the direction of the ambient geomagnetic field. Mantle convective motions cause the oceanic crust to move laterally away from the crest, giving rise to a 'magnetic tape recording' of changes in the polarity of the geomagnetic field (indicated by black and white blocks of crust). Both the heat-flow and topography decrease systematically away from the crest, but the gravity field and thickness of oceanic crust remains approximately constant, except within the central rift valley. On this basis the topography must be compensated by lower density mantle rocks, the asthenosphere, where their greater temperature results in a lower density. The igneous crust is covered by a gradually increasing thickness of sediments away from the ridge and the age of the basal sediments also increases away from the central crest. Hydrothermal circulation systems, often metal rich, occur close to the crest from which ores may be deposited when these are cooled by sea water penetrating the crust. Volcanogenic ores are also associated with the magmatic activity at the ridge crest (Chapter 6). (The scale for the thickness of sediments is exaggerated as these are up to 1-2 km thick, while the asthenosphere increase thickness from zero to a maximum of about 120 km by some 150-600 km from the ridge crest.)

Introduction: Plate Tectonics

Fig. 1.3. A model of a subduction zone. The descent of oceanic lithosphere is accompanied by seismic activity. The earthquakes are predominantly tensional (T) at the crest of the lithosphere as it bends near the trench, but compressional (C) at slightly greater depths. Earthquakes also result from shear (S) between the continental (island arc) and oceanic lithosphere but most deeper earthquakes are of double-couple (D) or uncertain (?) origin. The dehydration of the descending lithosphere gives rise to magmatic activity in the mantle overlying it. The composition of these resultant volcanics varies systematically with the depth of earthquake activity associated with the descending lithosphere, for example K : Na (normalised for Si content) and the rare earth elements (REE). These variations in volcanic composition are also mirrored in a systematic change in the composition of the metal ores (Fe, Mn, Cn, Pb, Zn, Au, Ag) associated with the volcanic processes (Chapter 7). The heat flow, and nature of the tectonic processes, varies from one area to the next, and the potential for hydrocarbon accumulation correspondingly varies (section 3.2).

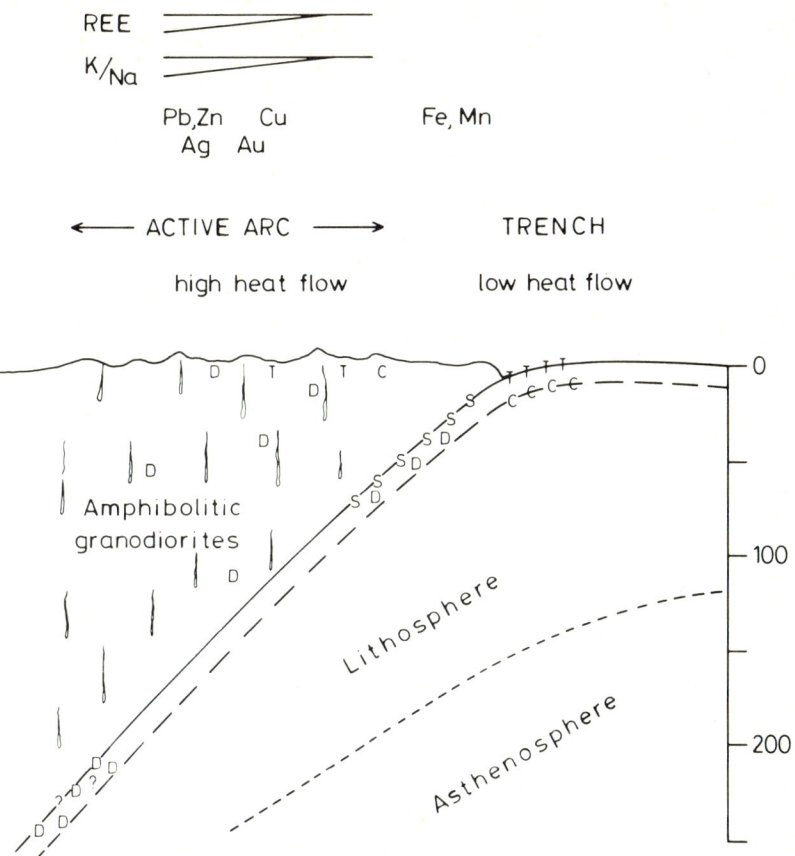

of subduction zones probably changes more rapidly than that of ridges as continental crustal rocks entering such zones would be too light to be carried into the mantle. Hence the arrival of continental blocks at a subduction zone restricts significant further subduction and consequently extant subduction zones must become more active or new subduction zones must be initiated following the blockage of subduction in any one area. A further consequence of the inability of continental crustal rocks to be subducted is that these blocks would come into collision at subduction zones. Such collisional orogenies may be of two previously separate continental blocks, as for the Alps and Himalayas, two island arc systems, as in parts of Southeast Asia, or continental blocks and island arcs, as in Western Canada.

The plate tectonic model is thus capable of providing an explanation for many past and present tectonic features of the Earth. It is consistent with the present knowledge of the age of the oceanic floors, all of which appear to be less than 200 million years old, and with the evidence for the previous contiguity of the world's continental blocks. In the context of the concentration of fossil fuels and ore deposits the theory must have many important consequences. The foremost is that the rate at which the Earth's internal heat is carried to the surface is closely linked with plate boundaries. The highest heat-flows occur along constructive boundaries and in the island arcs overlying subduction zones. The lowest heat-flows are associated with the oldest parts of the continental blocks and the oldest parts of the oceanic plates. Volcanogenic ores are necessarily associated with areas of high heat-flow, so their past occurrence must have been directly related to the previous location of plate boundaries. In general, such high heat-flow areas are inimicable to the preservation of hydrocarbons and thus such locations are likely to be inversely related to the occurrence of major oil and gas fields. The fact that the Earth's surface is mobile also means that past geographies must have differed drastically from today. Different areas would also have been in different latitudes at different times. Such considerations are of obvious importance with respect to the formation of different sedimentary facies, their organic content and hence their associated hydrocarbons, coal and metallic ores.

The succeeding sections of this chapter will consider the relevant properties of different plate margins, as recognised today; the possible evolution of such boundaries during the last 200-300 million years; the forces which appear to be responsible for such motions, and finally an assessment of the probability of such processes operating in even earlier times.

1.2 Constructive (accretionary) plate margins and the oceanic lithosphere

The seismically active oceanic ridges of the world's ocean form a single continuous unit some 80 000 km long. It passes through the Siberian Arctic, the Norwegian-Greenland Basin, the Central and South Atlantic, the Indian Ocean, with one branch passing into the Gulf of Aden and Red Sea and the other passing, as the Pacific-Antarctic Ridge, between Antarctica and Australia-New Zealand, eventually forming the East Pacific Rise (Fig. 1.1). In the Southeastern Pacific, the East Pacific Rise is linked to two other ridges, the Galapagos and Cocos Ridges, before passing northward into the Gulf of California. It is assumed to continue beneath the Basin and Range Province of the western United States and then links to the Juan da Fuca Ridge off Vancouver, which eventually disappears beneath Alaska. This total ridge system is locally offset by faults but essentially comprises a single morphological and tectonic unit. Similar, but much smaller, apparently isolated ridge systems also occur within the Scotia Sea (Barker, 1972) and are suspected to lie within some of the western Pacific marginal basins (Weissel, 1980). The oceanic ridges are characteristically some 2-3 km higher than the bordering ocean basins

Introduction: Plate Tectonics

and are some hundreds of kilometres wide. The crest of the rise is marked by the occurrence of numerous shallow earthquakes and, in many areas, by the existence of a central rift valley which, in the North Atlantic, is typically some 2 km deep, 20 km wide but of variable length. The earthquakes along the axis of the ridge are constrained to the upper 10-15 km and are generally weak, less than 6 magnitude on the Richter scale (Francis & Porter, 1973). More energetic earthquake activity results from strike-slip motions along the transform faults between the offset ridge crests

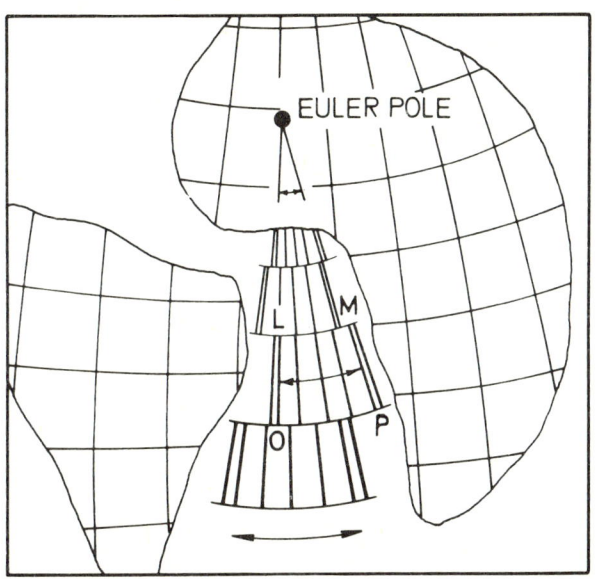

Fig. 1.4. Euler poles and rotations. The early opening of the South Atlantic can be readily defined as a pivoting rotation of South America away from Africa. The motion is about a point in the Sahara, a Euler Pole, the position of which can be defined by means of transform faults (these form concentric small circles around it) and magnetic anomalies (that broaden as they radiate away from it). The rate of the opening can be expressed as an angular opening in degrees per million years. For example, if the age of the points L and M are the same, then the summation of the opening rate can be defined readily and checked by other data, e.g. the age of O and P should be identical (after Tarling 1978).

(Fig. 1.4). Weak dip-slip earthquakes occur along the transform faults where they extend beyond the offset ridge crests. This activity arises from the different contraction rates as oceanic rocks of different age cool while adjacent to each other. The ridge crests are also characterised by very strong magnetic anomalies, with the magnetisation of the rocks enhancing the direction of the present geomagnetic field. Parallel with the ridge crests are magnetic anomalies that indicate that the crustal rocks below them are magnetised either in the direction of the present geomagnetic field, normally magnetised, or exactly opposite to it, reversely magnetised (Vine & Matthews, 1963; Morley, 1963). The sequence of magnetic polarities observed along traverses perpendicular to the ridge crests are identical to the known age sequence of polarity changes of the geomagnetic field during the last few million years (Cox et al., 1963; McDougall & Tarling, 1963). The parallel magnetic anomalies are associated with near surface rocks which cooled in the geomagnetic field at the time of their emplacement and hence acquiring that direction (Fig. 1.2). As the oceanic plates separated and new oceanic rocks were created, these newer rocks would acquire a magnetisation in the direction of a later ambient field. If the field had reversed polarity these newer rocks would thus be magnetised anti-parallel to the previous geomagnetic field. As the polarity sequence of the geomagnetic field is now fairly well known, the magnetic anomalies can be used to establish the age of the igneous rocks of the ocean floor formed during the last 200 million years. The age sequence derived magnetically is the same as that indicated by the age of the oldest sediments overlying the igneous rocks of the ocean floor. Where these have been penetrated by core sampling or drilling, the age of the oldest sediments increases systematically away from the crest of the oceanic ridges in exactly the same way as the age of the igneous rocks as dated palaeomagnetically (Fig. 1.2).

The oceanic ridge crests are characterised by overt volcanism in some areas, such as Iceland and Tristan da Cunha. Submarine studies of the ridge crests are naturally few but detailed studies, such as those in the North Atlantic rift valley near the Azores (Moore *et al.*, 1974; Aumento *et al.*, 1975; Lilwall *et al.*, 1978; Fowler & Keen, 1979) indicate episodic volcanism taking place along the centre of the rift valleys along their entire length. The crests are thus associated with abnormally high heat-flow (Fig. 1.2). It is, in fact, difficult to make such geothermal flux determinations near the crest of the ridge because the measurements are conventionally taken using a core barrel to penetrate sediments within which the temperature is measured at different depths. The thermal conductivity of the intervening sediments is then determined from samples within the core barrel (Bullard, 1954). These measurements are, of necessity, only taken in sedimentary pockets but these are scarce on the crest of the oceanic rises and those that do occur are likely to be in atypical localities. They would, for example, generally occur where tensional faulting had created a sediment trap and are thus likely to be directly associated with an underlying fault zone. There are also some fundamental problems interpreting such observations as this method assumes that the Earth's internal heat is being transferred to the surface by thermal conduction alone yet it seems probable that hydrothermal circulation systems are likely to be even more effective agents for heat transference in these areas. On this basis it seems likely that many of the heat-flow estimates for ridge crests are likely to be significant underestimates of the total heat being transported to the Earth's surface in such localities. Nonetheless it is clear that there is an intimate association between accretionary, constructive boundaries and abnormally high geothermal high heat-flows. Both heat-flow and topography decrease systematically away from the oceanic ridge crests; both are dependent on the square root of the age of the oceanic igneous rock floor and overlying sediments (Goslin *et al.*, 1972; Parker & Oldenburgh, 1973; Davis & Lister, 1974; Kono & Yoshii, 1975; Sleep, 1975; Yoshii, 1975). There are, of course, departures from such general behaviour with local, high heat-flow spots being apparently associated with the circulation of hydrothermal fluids through at least the upper part of the oceanic crust (Bodvarsson & Lowell, 1972; Bottinga, 1974; Lister, 1974; Davis & Lister, 1977; Langseth *et al.*, 1977, Epp & Suyenga, 1978). Some of these hot spots also appear to be associated with metallogenesis at the igneous-sedimentary contact, particularly at distances of some 5-10 km away from the centre of the ridge crest (Hart & Staudigel, 1978).

Refraction seismic surveys of the oceanic crust indicated that it is remarkably uniform in composition and thickness throughout all of the ocean basins (Bott, 1971; Cann & Moore, 1978; White, 1979). Four specific layers are almost always identifiable, with the upper layer comprising sediments (Cann, 1974; Kusznir & Bott, 1976; Tarling, 1978a). The second (sub-sediment) layer is generally some 1½ km thick and comprises basaltic lavas and dykes which overlie layer 3, the exact composition of which is not fully established but seems to comprise gabbroic rocks, some 5 km thick, that are predominantly in an amphibolitic grade of metamorphism. The bottom layer, 4, has only been recognised within the last few years (Cann, 1974; Christensen & Salisbury, 1975) and appears to be thin, less than 1 km thick, and comprises an ultrabasic, layered complex directly overlying the mantle rocks. It is generally thought that such crust forms by differentiation processes of mantle rocks at the crest of the seismic oceanic ridges (Chapter 6.2). Layer 3, may, in fact, be formed by the cooled walls of a magma chamber which has differentiated upwards to form the basaltic layer and downwards to form the layered complex (Whitmarsh, 1975). However, there is some doubt as to the actual existence of such magma chambers as these are not clearly identifiable on seismic profiles along the crest of the Atlantic or Pacific ridges. It seems more probable that such differentiation mainly takes place in cracks within layer 3, rather than within a discrete magma chamber (Arzi, 1978; Nisbet & Fowler, 1978). The degree of symmetry of the magnetic anomalies bordering the crest of the oceanic ridges suggests that the volcanism, associated with such magma differentiation, is episodic on a scale of a few 10 000 years, but is essentially continuous when considered on a time scale of 2-3 million years (Ramberg *et al.*, 1977).

The whole oceanic ridge is in gravitational (isostatic) equilibrium, except for very local features associated with the 5-10 km area bordering the seismically active central zone (Fig. 1.2). As the oceanic crust is so uniform in composition and thickness, the gravitational effect of the topography of the ridges must be compensated for by lower densities within the mantle. The volume of such low density mantle must thus decrease according to age$^{1/2}$ of the overlying crust in the same way as the topography. It is generally considered that this decrease in density is almost, possibly entirely, due to the expansion of mantle rocks due to excess heat within them and that this heat is systematically lost as the oceanic lithosphere migrates away from the ridge (Green & Lieberman, 1976; Froidevaux *et al.*, 1977).

The motion of the oceanic lithosphere is clearly recorded by means of the magnetic anomalies (Blakely & Cande, 1979) and these record the time at which that part of the igneous oceanic crust was acquiring its magnetisation as it cooled from molten temperatures at the ridge crest. The offset of the anomalies also allows a direct measurement of the offset of the ridge crests along a fault (Fig. 1.3). In general such transform fault zones (Wilson, 1965; De Long *et al.*, 1979) are relatively narrow but if the direction of spreading on different sides of a transform fault exceeds a few degrees, then the transform fault appears to act as a weak area through which mantle rocks penetrate. The crustal growth is irregular so that as they cool their corresponding magnetic anomalies are also irregular.

There are areas where three plates are contiguous so that a triple junction (Fig. 1.5) exists that is geometrically and mechanically stable (McKenzie & Morgan, 1969). Such triple junctions may comprise of various combinations of rift, transforms or subduction zone, although those incorporating a subduction zone are not stable over long periods of time and the relative configuration must change significantly as the system evolves. The direction and rate of motion of the plates, relative to each other, can be determined by using the shape of the magnetic anomaly patterns or transform faults. The magnetic anomaly pattern must clearly radiate away from a Euler pole (Fig. 1.4)—the pole about which the motion of the plates can be defined—and transform faults must form small circles concentric to this pole. The rate of relative rotation can then be defined in terms of the rate of rotation of the individual plates about this Euler pole. The motion of any plate is therefore defined precisely by means of the latitude and longitude of its Euler pole and the rate of rotation (usually in degrees per million years) about that pole (Le Pichon *et al.*, 1973). Most plates show changes in the rate of rotation with time, even with a fixed Euler pole position, but there are also times when the position of the Euler pole changes. In the case (Fig. 1.4) of the opening of the South Atlantic (Tarling, 1980a), the original opening took place relative to the Euler pole in the vicinity of West Africa so that South America initially appeared to pivot away from Africa, but remained

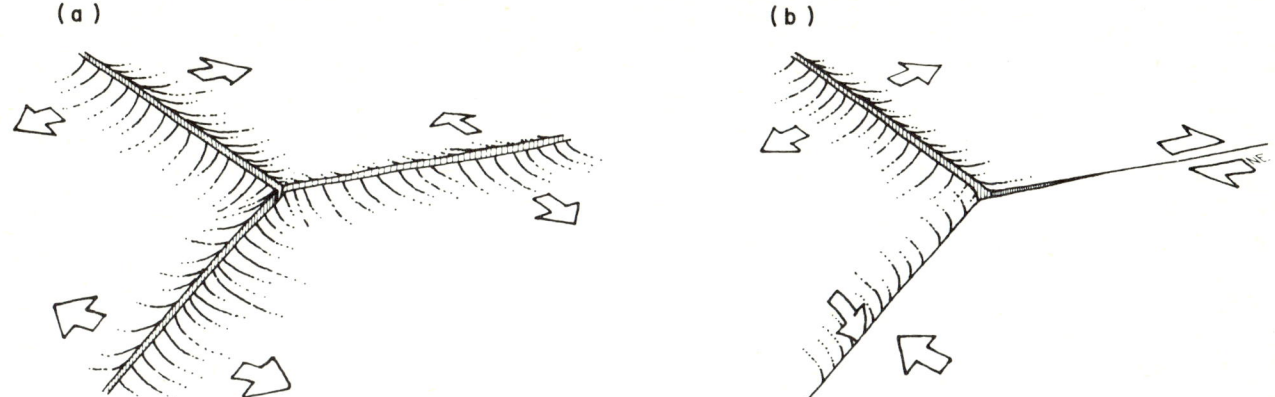

Fig. 1.5. Triple junctions. The three types of plate boundaries (oceanic ridges, subduction zones and transform faults) can occur singly, pass from one to the other, or link in various combinations of all three. Six stable combinations of triple junctions exist today (McKenzie & Morgan, 1969) of which two are illustrated here. (a) A junction of three ridges, as at the East Pacific Rise - Galapagos Rise intersection; (b) a triple junction involving each type of boundary, as off the Gulf of California. Ancient triple junctions when partially located on continents, may be particularly important economically. For example, where one of the rift junctions was unable to open fully (a 'failed' arm), the resultant rift valley is frequently the site of metallogenesis, association with initial hydrothermal and volcanic activity, and subsequently hydrocarbons that may be generated from organic debris as they infill such sedimentary traps. (Drawn by N. B. Eustance.)

contiguous between Brazil and West Africa, but such a motion became essential east-west during the last 80 million years or so when the South Atlantic Euler pole became positioned close to the Earth's axis of rotation.

1.3 Destructive plate boundaries (subduction zones) and the continental lithosphere

On reaching a subduction zone, the oceanic lithosphere plate appears to be strongly bent and its upper surface has associated tensional seismic activity (Sacks *et al.*, 1978), while earthquakes deeper within the plate appear to be primarily compressional (Fig. 1.3). The strongest seismic activity occurs as the oceanic lithosphere actually descends into the Earth's mantle along an inclined plane, the Benioff zone (Benioff, 1949). Most of these earthquakes, down as far as 600-700 km, appear to be confined within a 10-12 km band and may thus be confined within the descending oceanic crust. In some areas, the Benioff zone may comprise two even narrower zones (Engdahl & Scholz, 1977; Hasegawa *et al.*, 1978a, b, 1979; Yoshii, 1979). Intermediate depth and shallow earthquakes also occur in the area overlying the dipping Benioff zone. These appear to be mostly associated with the rise of magmas from near the vicinity of the Benioff zone. These magmas eventually feed the volcanic activity within the island arc or orogenic belt bordering the trench.

The general interpretation of these zones is that the process of subduction of the oceanic lithosphere gives rise to major stress within it, due to both the increased pressure and temperature, and also to physical reactions with the mantle rocks beneath which it is passing. The specific interpretation of how magmas are generated is variable. On some models it is suggested that the descending oceanic crust becomes partially molten and gives rise to the ascending magmas (Miyashiro, 1972). Such a model is not necessarily inconsistent with the geochemical evidence. For example, strontium isotopes indicate a mantle derivation for the igneous rocks of the island arcs (Ringwood, 1975), although such derivations would also be consistent with their formation from oceanic rocks because these were formed from mantle rocks themselves, at the oldest only 200 million years before. However, it seems more probable that the magmas are generated from the mantle overlying the descending lithosphere (Ringwood, 1974, 1975; Taylor, 1979). As the oceanic rocks descend, water is expelled from them as a result of both temperature and pressure increases. The rise of such water, with other volatiles, into the overlying mantle rocks lowers the melting point of the mantle rocks and so partial melting ensues. The magmas generated then migrate towards the surface (Marsh & Kantha, 1978) but show some geochemical zonation (Kuno, 1959, 1966; Dickinson, 1970, 1975; Jakes & Gill, 1970; Jakes & White, 1972; Miyashiro, 1975; Hutchinson, 1976; Francis *et al.*, 1977; Sakuyama, 1977; Moorbath *et al.*, 1978; Arkani-Hamed, 1978) reflecting the difference in the pressure and temperature of the zones from which they originated and also the thickness of mantle rocks through which they have passed before being erupted at the surface. Thus there is an increase in K/Na (normalised for Si content), the rare earth elements, particularly the lanthide elements, and also a gradation of the composition of the metallic ores (Mitchell, 1973; Sillitoe, 1972; Tarling, 1973)—all associated with different depths of origin of the parent magmas (Fig. 1.3 and Chapter 7). It is also clear that, under such conditions, there will be radically different pressure-temperature conditions in the top few kilometres of the Earth's surface. Blueschist metamorphism tends to characterise the rocks in the immediate neighbourhood of the oceanic trench because the pressures are high yet the temperature is low. Over the Benioff zone, the pressures are similarly high, but the temperature is much higher and the metamorphism is correspondingly more intense, giving rise to 'paired' metamorphic belts in which zones of blueschist (pumpyllite-prehenite association) and amphibolite grade metamorphism occur parallel to the subduction zone (Miyashiro, 1961, 1972; Mitchell, 1974; Ernst, 1975). These zones also provide a means for recognising the polarity of the subduction, i.e. in the case of Japan the subduction is taking place from the Pacific towards the Asian mainland. The recognition of similar paired metamorphic belts, as in the Alps, can therefore be used to determine the polarity of ancient subduction patterns. In the Alpine case the subduction appears to have taken place predominantly from north to south in the Switzerland-Austrian area, although obviously the total story is much more complex (Frisch, 1979).

The average composition of the volcanics produced in the island arcs overlying the Benioff zone is essentially the same as the average composition of 'normal' continental crust (Veizer, 1974, 1976; Veizer & Compston,

Introduction: Plate Tectonics

1976; Gunn, 1976; Taylor, 1979; Taylor & McLennan, 1979). Such similarities naturally lead to the conclusion that continental crust is being created today by the igneous differentiation of mantle rocks, and possibly of oceanic lithosphere, over the present Benioff zones. Such an explanation seems inevitable when areas are considered, such as along the western Aleutian Trench, where 'continental' calc-alkaline igneous rocks have been generated by the subduction of oceanic lithosphere beneath oceanic lithosphere (Coats, 1962).

Continental crust is generally 20-30 km thick, but may reach thicknesses of 60-70 km beneath presently active orogenic belts. It is sometimes characterised by a predominantly two-layer but very complex structure, with a lower crust of higher density. It is also thought that the majority of continental crust, particularly the lower crust, is in a granulite grade of metamorphism (Tarney & Windley, 1977; Smithson, 1978; Hall & Simmons, 1979) while amphibolitic grade seems more characteristic of island arc situations. The higher density of the lower crust can not be attributed solely to the different metamorphic grade but probably also reflects the presence of a greater number of basic igneous intrusive bodies. However, its main characteristic is its complexity (Smithson *et al.*, 1977; Schilt *et al.*, 1978; Smithson, 1978). This may well reflect its much longer history as the oldest rocks approach some 4000 million years, in contrast to the oceanic crust of age 200-250 million years. The older, Archaean, rocks form relatively small cratonic blocks and are characterised by two main types of terrain (Windley, 1976): the greenstones, which form 15-20% of the Archaean rocks, and the granite-gneisses that form the majority of this age crust. Both terrains appear to have been derived from the mantle because both have mantle-type isotopic ratios (Moorbath, 1976; Lambert *et al.*, 1976; Gunn, 1976). The granite-gneisses are often interpreted as 'continental' rocks and the greenstones as 'oceanic'. However, the relationship between the two terrains is still unclear and it seems probable that the Archaean greenstones are, in fact, more equivalent to Mesozoic continental flood basalts, such as those of the Triassic Siberian Traps and the Triassic-Jurassic Karroo basalts of southern Africa. The rocks of younger Precambrian age, the Proterozoic, date between 2 500-2 700 million years and the start of the Cambrian age, 570 million years ago. These rocks indicate large-scale stability of the continental crust during this time and the greenstones of Archaean times are largely absent (Sutton & Watson, 1974).

The tectonics in the Phanerozoic do not differ significantly from those of the Late Proterozoic. There is clear evidence that continental splitting and sea-floor spreading occurred during at least the last 200 million years, and the Caledonian-Appalachian orogenic belts can also be explained in

Chapter 1

similar terms. The occurrence of continental splitting and sea-floor spreading in earlier times is not so clear (Dewey & Spall, 1975). Such mechanisms could explain the Grenvillean orogeny some 1 000 million years ago (Irving, *et al.*, 1974), and it is also arguable that similar processes were in operation in the Huronian (Shaw, 1976) and even for the Coronation geosynclinal evolution in the early Proterozoic (Hoffman, 1973). While the evidence for such tectonics is unclear and somewhat subjective there seem to be major differences between the tectonic patterns of today and those of the past, but further discussion of these will be deferred (section 1.6) until after consideration of the nature of the mechanisms that operate today.

One of the major features of the continental crust is the relationship between geothermal heat flow and the age of the last orogenic event in any one region (Chapman & Pollack, 1975; Kono & Amano, 1978; Allis, 1979). The relationship depends, as in the case of the oceanic crust, on the age$^{1/2}$ of the last orogenic event, but the continental time scales are very much greater. This relationship is extremely important as it apparently conflicts with surface evidence for the distribution for radiogenic, heat-producing elements. These are mostly associated with the exposed Precambrian granites. On this basis it would be expected that such areas would, in fact, have higher heat-flow than younger areas. As the converse appears to be true, it means that the radiogenic elements must be strongly concentrated into the surface levels of the continental crust and that there is a corresponding depletion in radiogenic-heat producing elements in the lower crust and also within the upper 200-300 km of the mantle (Pollack & Chapman, 1977; Allis, 1979). Such an assessment clearly means that this thickness of mantle, at least, must move with the continental crust and thus the continental lithosphere must be at least 200-300 km thick. Such an observation appears to be confirmed by a range of other indicators (Jordan, 1975; Jordan & Fyfe, 1976; Mori & Green, 1976; Osmaston, 1977). There is no evidence for the existence of a seismic low-velocity layer in the mantle which underlies the Archaean blocks (Poupinet, 1979), for example that in eastern Canada (Jordan & Frazer, 1975) and western Australia (Gonz & Cleary, 1976). Similarly the occurrence of kimberlites appears to be confined to Archaean cratons, although themselves of much younger age, e.g. some South African kimberlites are of Cretaceous age. Knowing that the diamondiferous kimberlites, at least, must have originated at depths of some 150-200 km (Dawson, 1977; Meyer, 1979), it is again clear that this thickness of mantle must move as a unit with the overlying Archaean rocks and therefore that the continental lithospheres must, as a minimum, be of this thickness beneath the older cratonic blocks (Jordan, 1975; Pollack &

Chapman, 1977). Indeed, it is difficult to see any seismic distinction between the base of the continental lithosphere and the 'rigid' mantle rocks that lie beneath the seismic low-velocity zone in the oceanic regions.

1.4 Continental separation during the last 300 million years

It seems convenient to restrict this review to the last 300 million years because it is during this time there is strong evidence for differential movements of continental blocks from oceanic anomaly patterns as well as palaeontological, palaeoclimatic and continental palaeomagnetic evidence (Tarling & Tarling, 1975). For earlier times, continental palaeomagnetic and palaeontological evidence is available for the Palaeozoic, but there are no data from oceanic anomalies. The end of the Carboniferous and Early Permian is also the time when most of our present continents were in reasonably close proximity. The 'southern' continents, South America, Africa, Madagascar, India, Australia, New Zealand and Antarctica, were still united as the supercontinent of Gondwanaland. North America and Europe had linked to form Laurentia at the end of the Caledonian orogeny, and this supercontinent was in the process of linking with the Siberian platform. Most, but not all, continents were therefore together forming a single continental unit, called Pangaea (Fig. 1.6). Much of southeastern Asia, particularly the Chinese blocks, were not part of this unit. Laurentia and Gondwanaland, although in very close proximity, were moving relative to each other and were not, therefore, part of the same lithospheric plate (Tarling, 1979, 1980b). In fact the Late Carboniferous collision between northern South America and southern North America (Ross, 1979), at the site of the present Gulf of Mexico, was one of the most important features of the Phanerozoic. It appears to have reversed the direction of rotation of Laurentia and drastically changed the rate of rotation of Gondwanaland (Tarling, 1979). It seems probable that the impact was also sufficient to cause extensive fracturing of both supercontinents during the Permian as most areas, destined to become continental edges now bordering the Atlantic and Indian Oceans, were faulted at this time (Kent, 1976, 1977).

The Antarctic section of Gondwanaland then lay in southern polar regions, but only high ground was subjected to glaciation during the Late Carboniferous. The Gondwana ice sheet (Tarling, 1978b), affecting all Gondwanan continents, was not initiated until the Sakmarian stage of the Permian (Anderson & Schwyzer, 1978). This extensive ice sheet disappeared from all these areas by the end of this stage, leaving only mountain glaciation in areas of high ground in polar latitudes. While Gondwanaland mostly lay in high southern latitudes, most of Laurentia lay in equatorial latitudes, although a marked change in the palaeo-

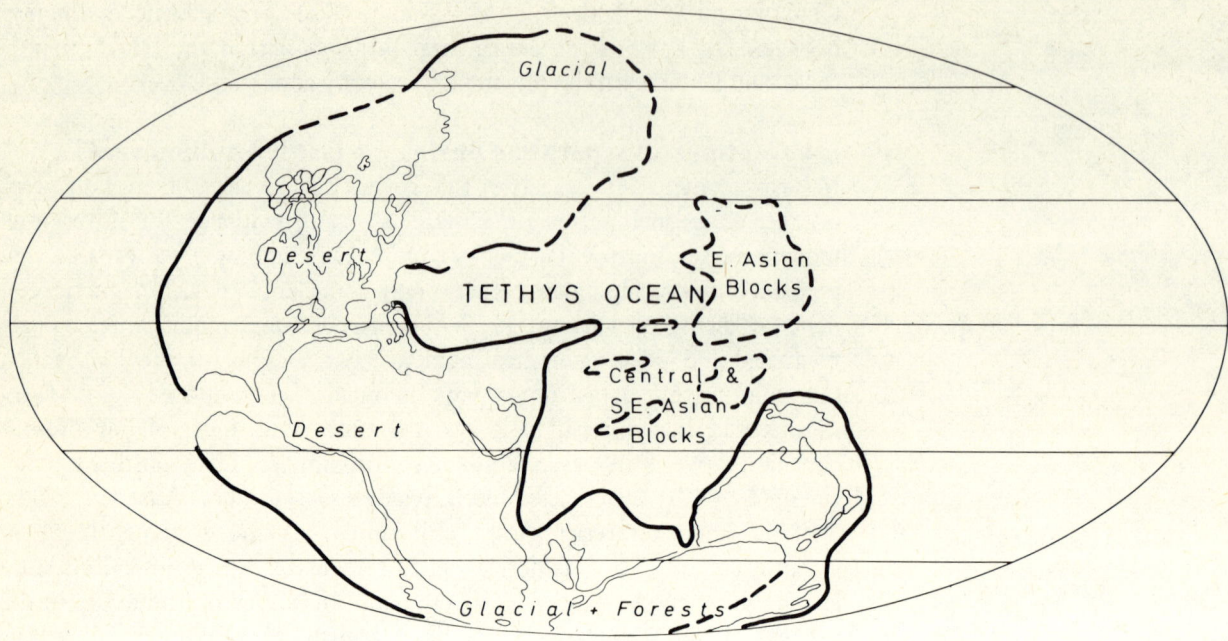

Fig. 1.6. A Permian palaeogeographic reconstruction (c. 250 million years ago). This reconstruction is based primarily on palaeomagnetic data but is broadly consistent with the palaeoclimatic evidence for the Lower Permian glaciation in the southern continents—Gondwana—and at slightly later, Upper Permian glaciation in Siberia. The Gondwanan glacials were followed by high-latitude fern forests, that eventually formed the Gondwanan coal deposits, but the equatorial rain forests of Europe and North America that formed the coals of Late Carboniferous age had then given way to low latitude desert sandstones. The precise location of China, various southeastern and central Asian blocks is still unclear at this time.

geography and climate had resulted in the cessation of the equatorial fern forests (Chapter 5) towards the end of the Carboniferous. The Chinese and southeastern Asian continental blocks appear to have laid in similarly low latitudes (Ridd, 1971; Kremp, 1977), as they had through most of the Palaeozoic. The Siberian block was in high northern latitudes, where it underwent extensive glaciation in the Upper Permian prior to its collision and suturing on to the Laurentian land mass along the line of the Ural Mountains at the end of the Permian. Meanwhile, the differential motion between Laurentia and Gondwanaland had resulted in yet another closure along the line of the Central Atlantic, forming the Variscan Mountains in Early Permian times.

The subsequent split of this Pangaean supercontinent occurred only shortly after the addition of the Siberian block. The central Atlantic, between Africa and North America, began to form in earliest Jurassic times, some 180 million years ago. This split seems to have been associated with the development of a spreading ridge system that passed through the present Gulf of Mexico and was probably continuous with ridge systems in the Pacific Ocean (Tarling, 1980a). The initial split within Gondwanaland appears to have been between the western side of India and Africa-Madagascar and only shortly afterwards, possibly in the mid Jurassic, some 160 million years ago, with associated extensive flood-basalt volcanism of the Rajmahal Traps, the Karroo, Tasmanian and Serrar Basalts in India, Southern Africa, Australia and Antarctica, respectively.

Introduction: Plate Tectonics

The volcanism in Antarctica and Australia was possibly more closely related to movements between Eastern and Western Antarctica although such motions are hypothetical and necessitated by difficulties in reconstructing Gondwanaland. The separation of India from Africa-Madagascar may have been contemporaneous with, or slightly earlier than, the separation of India from Antarctica. Thus India became surrounded by shallow marine seas by the Late Jurassic and became increasingly isolated as these marine conditions expanded into deep oceanic areas. As India separated from Antarctica, Antarctica also began to separate from Africa, although it remained linked to South America along the line of the southern Patagonian Andes and also with Australia and New Zealand.

The Tasman Sea began to form in the Late Cretaceous, some 80 million years ago (Fig. 1.7), separating New Zealand from Antarctica-Australia, although this separation ceased in the Eocene, some 45 million years ago, as Australia began to separate from Antarctica. The initiation of the Tasman Sea was therefore approximately contemporaneous with the final separation of Africa from South America. South America had been pivoting away from Africa, with a Euler pole in West Africa, commencing in very late Jurassic-basal Cretaceous times some 130 million years ago. This Euler pole location meant that West Africa and Brazil remained close together at this time. However some 80 million years ago, the location of the Euler pole for the Africa-South America motions moved to near the North Pole so that South America and Africa then separated in an east-west direction. Shallow marine connections had, however, been established between the Central and South Atlantic Oceans in the Turonian, some 91 million years ago (Reyment, 1969). The actual separation of Antarctica from South America appears to have been late, with the final break in the direct terrestrial connection between the two continents taking place in Oligocene times, with the initiation of full oceanic conditions taking place only 27 million years ago (Barker & Burell, 1977).

In the northern continents North America and Europe still behaved as a single unit, Laurentia, as they separated from Africa. Even further northwards, the Kolyma Block, then filling much of the Canadian Arctic Basin, began to separate in Late Jurassic or early Cretaceous times, eventually colliding with the Siberian Platform, forming the Cretaceous Verkhoyansk Mountains at the same time as several continental and thick oceanic units combined to form Alaska. In Mid Cretaceous times, c. 100 million years ago, (Fig. 1.7) a slight separation was initiated between North America and Greenland-Europe as the Labrador Sea-Baffin Bay area (Srivistava, 1979) began to form, accompanied by transform motion along the Nares Straits (Newman & Falconer, 1978). This opening was, however, small and terminated when sea-floor spreading started in the

Chapter 1

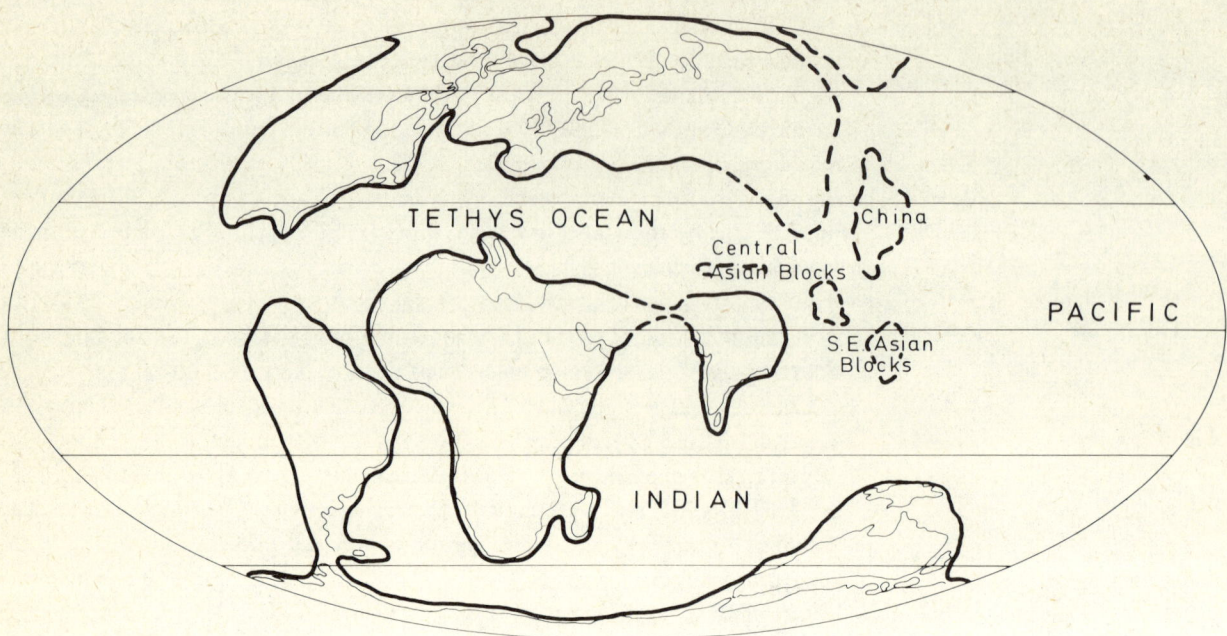

Fig. 1.7. A Cretaceous palaeogeographic reconstruction (c. 100 million years ago). This reconstruction is based largely on oceanic magnetic anomalies, supplemented by continental palaeomagnetic data. Much of Europe, central North America and Australia were covered by seas at this time. Major chalk formation in Europe probably reflected the mixture of cold arctic waters with warm Gulf Stream waters, resulting in prolific micro-organic growth reflected by the rapid deposition of pure carbonate shells and skeletal remains at this time (Hart & Tarling, 1968).

Norwegian-Greenland Sea area and along the northern edge of the Siberian block in Palaeo-Eocene times, some 50 million years ago. In the Arctic this spreading opened up the Siberian Arctic Basin and carried the Lomonosov Ridge off from its previous position along the edge of Siberia, to its present trans-polar location. In the Norwegian Basin the position of the spreading ridge 'jumped' at least five times during the evolution of that basin—one of the 'jumps' separating the continental Jan Mayen ridge from Greenland (Talwani & Eldholm, 1974). Similar types of jump were presumably associated with the separation of the Rockall Bank from Europe, the Blake Plateau from eastern North America, etc.

The most recent initiation of spreading has taken place in two areas; the Gulf of Aden-Red Sea, which commenced some 30 million years ago, ceased 15 million years ago, and recommenced during the last 5 million years (Girdler & Styles, 1978), and in the Gulf of California where Baja California has separated from the mainland since mid Miocene times.

1.5 The mechanism of plate tectonics

A variety of mechanisms for the motion of the continents have been invoked during the 20th century (Schmidt, 1978). Mechanisms originating externally to the Earth can now be largely or completely discounted on the basis that they would also operate on other planets, yet these (Mutt, 1979) only show evidence for very slight contraction since the formation of their lithospheres some 4 550 million years ago (McElhinny *et al.*, 1978). Possible mechanisms that are internal to the Earth fall into two basic categories: those that are confined within the plates themselves, with mantle motions responding to movements of the plates, and those in which plate movements arise in response to mantle motions.

There are two basic forms of intraplate mechanism (Richardson *et al.*, 1979). The first arises from the fact that the seismically active oceanic ridges are some 2-3 km higher than the surrounding oceanic basins so that there is a component of gravitational acceleration acting within the plane of the plate (Hales, 1969). This force is comparable with the total seismic and volcanic release of energy (Forsyth & Uyeda, 1975) but seems inadequate to explain all the tectonic features observed on the plates even if these are completely decoupled from the underlying mantle. The total tectonic pattern clearly necessitates a greater force than this. It seems probable that tensional faulting within mid-oceanic rift valleys may be associated with such forces as these would be particularly important on a local scale (MacDonald & Atwater, 1978) where the lithosphere is still extremely thin. The second major intraplate force arises because the oceanic lithosphere is cooling and hence it is becoming more dense away from the oceanic ridges. Its buoyancy, relative to the warmer underlying mantle rocks, thus becomes less as it becomes older (Molnar & Atwater, 1978; Molnar *et al.*, 1979). The main effect of such coolness is that during subduction phase changes, induced by increased pressure at greater depths within the mantle (Ringwood, 1958; Anderson, 1967, 1975; Press, 1971; Forsyth & Uyeda, 1975; Chapple & Tullis, 1977), will clearly take place at shallower depths within the cooler denser subducting lithosphere than in the 'normal' mantle at the same pressure on either side of it. These phase changes, which occur predominantly at 320 and 640 km depths, provide a major increase in the net density of the subducting slab and hence a strong downward force (Elsasser, 1969; Ringwood, 1976). This force, if coupled to the rest of the oceanic lithosphere, appears to be adequate to cause plate motions if the lithospheric plates are decoupled from the underlying mantle (Schubert & Turcotte, 1971). While such forces undoubtedly exist in subduction zones, there is still uncertainty about the effects of phase transitions on the process of motion (Sammis, 1976; Sammis *et al.*, 1977). There is also seismic evidence that the descending oceanic lithosphere is broken into discrete slabs as it descends (Baranzagi *et al.*, 1973, Liu, 1975) and such slabs cannot thus be coupled to the lithosphere within the oceanic basins. In any case it seems impossible for such a mechanism to be the main driving force as:

1 it requires subduction to be already initiated (in the same way as the ridge crest force requires the crest to have been initiated),

2 it cannot account for the presence of the Arctic, Atlantic and West Indian Oceans, as they have no associated subduction zones,

3 there is no evidence from the magnetic anomaly patterns that there is any change in the rate of ocean floor spreading when the oceanic lithosphere reaches the different depths at which major phase transitions take place,

4 'jumps' or pauses in the locus of spreading can be explained more simply by changes in the locus of the rising convection cell than by changes in the forces at the ridge or trench,

5 the system involved is necessarily a convective system involving return flow and it is not possible to talk separately of 'push' or 'pull' mechanisms as both push and pull must act simultaneously.

From such considerations it seems most unlikely that either intraplate mechanism is the fundamental driving force for plate tectonics, although clearly such forces are extremely important when considering tectonic processes at oceanic ridges and subduction zones.

The most likely situation (Fig. 1.8) appears to be similar to that proposed by Holmes (1927, 1944) and Daly (1940). Motions of the surface of the Earth occur in response to mantle motions driven by the heat released by radioactive disintegration of potassium, thorium and uranium within the mantle. The thickness of the continental lithosphere and the absence of a seismic low-velocity zone beneath the cratonic blocks (section 1.3) means that such mantle motions cannot be restricted to the seismic low-velocity zone, as in conventional models (Isacks *et al.*, 1968; Le Pichon *et al.*, 1973; Schubert *et al.*, 1978), but must extend to depths much greater than 200 km. As the seismic evidence indicates a uniform composition for the mantle below some 200 km (Birch, 1952), and this also appears to have asthenospheric viscosities of some 10^{21-22} Poise (Cathles, 1975), it seems reasonable to assume that such mantle motions can extend to within 200 km of the Earth's core (Runcorn, 1972; Tozer, 1977).

Such a conclusion means that the conventional interpretation of the nature of the seismic low-velocity zone must be seriously wrong. This zone is generally interpreted as an area within the mantle where the geothermal gradient and the mantle solidus are so close to each other that some 0.1% partial melting occurs (Ringwood, 1975). The molten fraction thus accounts for the attenuation and delay of seismic waves passing through this zone (Solomon, 1976). If this zone really is of partial melt, then there is clearly a drastic change in rigidity between it and the overlying lithosphere. It is also likely that the overlying lithosphere would

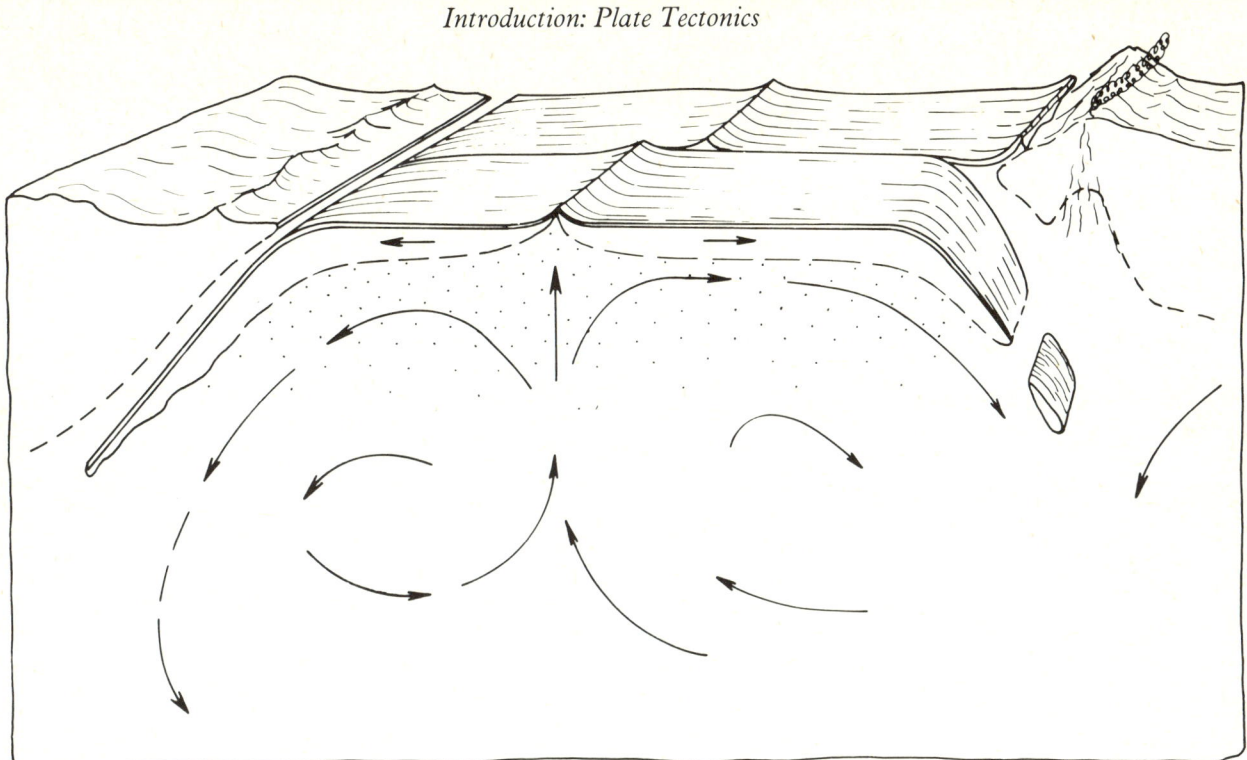

Fig.1.8. The driving mechanism of plate tectonics. Mantle-wide convective motions cause sea-floor spreading and subduction, with only local tectonic effects arising from intraplate features. The lithosphere is strongly coupled to the asthenosphere which, coincidentally, corresponds with the seismic low-velocity zone within the oceanic mantle regions. This correspondence arises because the formation of free intergranual water within the oceanic mantle takes place at approximately the same temperature at which the viscosity decreases to allow ready plastic deformation, i.e. some 200°C below the melting point of mantle rocks. The continental lithosphere is some 200-250 km thick, but also overlies convecting mantle rocks. These rocks are too deep to contain significant intergranular water and their overlying continental lithosphere has already been dehydrated during the evolution of the continental lithosphere. (After Tarling, 1978.)

be mechanically decoupled from the rocks of the seismic low-velocity zone—a feature that is a basic assumption in models which consider the intraplate forces to be the driving force of plate tectonics.

The seismic low-velocity layer may equally well represent a level in the mantle at which pressure-temperature conditions permit water to be present as a free phase, i.e. as an intergranular film. Such water could account for the observed seismic velocity properties, but would not necessarily mean the mantle had reached its melting temperature. The upper boundary of this layer is almost certainly an isotherm because its depth is directly related to the age, topography and heat-flow observed at the surface of the oceanic plate. However, such an isotherm could well be of the order of 800-900°C, and correspond to the dehydration of amphibolites—rather than 1100-1200°C melting temperature proposed by the previous explanation. A temperature of 800-900°C for the upper surface of the seismic low-velocity zone would, in fact, mean that this boundary would coincidentally correspond to the approximate base of the lithosphere if this is defined as occurring at the level of the decrease in viscosity at 200°C below the melting point of the mantle. The lower boundary of the seismic low-velocity zone is poorly defined but appears to be fairly uniform on a world-wide scale. It presumably corresponds to a pressure-induced reabsorption of water into normal mantle phases.

Chapter 1

Such an explanation for the nature of the lithosphere and asthenosphere is critical in assessing the temperature and, hence, conditions at which magma generation will occur. It is important to stress, however, that if the conventional model of partial melting is correct, then it is not valid to interpret the top of the seismic low-velocity layer as the base of the lithosphere as the relationship between viscosity and melting temperature means that major plastic deformation will take place at temperatures down to 200°C below melting point of the materials concerned (Vetter & Mesner, 1977; Pollack & Chapman, 1977). The base of the lithosphere must therefore be appreciably shallower than the level at which partial melting occurs.

1.6 Past plate tectonics

Any assessment of the nature of tectonics occurring during the Precambrian must be approached from two directions. The first is to consider the mechanism of plate tectonics today and determine:

1 whether such processes could have operated in the past
2 which geophysical or geochemical parameters may have been different in the past and hence caused such processes to operate in somewhat different ways

The second approach is to examine the geological evidence for evaluating the probable nature of the tectonic regime at different times during the Precambrian.

If the mechanism of present day plate tectonics is primarily driven by radiogenic heat produced within the mantle, then the fact that the amount of geothermal heat being produced in the past was higher than today would imply that mantle convection would have been more vigorous and, presumably, its surface expression would be similarly enhanced. The rate of convection is, to a first approximation, dependent on the heat production$^{1/2}$ within the mantle (Tozer, 1977). Most estimates of the concentration of radiogenic-heat producing elements suggest that the total radiogenic heat being produced in the Earth at the start of the Archaean, some 4 000 million years ago, was some five to six times greater than today (Lambert, 1976; Bickle, 1978). This suggests that net mantle convective rates were about twice as great as those today. However, conditions within the Earth would also have been different as the composition of the mantle, for example, incorporated those elements that were to be later differentiated into continental crust.

A major constraint on earlier conditions would have been the thinness of the lithosphere. The surface temperature of the Earth seems to have been essentially constant throughout almost all of the Earth's history—as atested by the presence of running water. Similarly, the temperature at the base of the lithosphere is, by definition, constant. The thickness of the lithosphere is thus defined by the geothermal gradient, or, more pragmatically, the geothermal gradient determines the thickness of the lithosphere. As the greater past heat production must be transferred to the Earth's surface by convection and then conduction, yet the temperature regime within the convecting mantle must remain constant, the surface geothermal gradient must have been greater, on average, and the average lithosphere must therefore have been thinner. One important consequence of the thinner lithosphere was that isostatic adjustments must have taken place on a smaller scale and more rapidly than today (Hargraves, 1976; Tarling, 1978a, 1980c). It seems probable that neither deep sedimentary troughs, nor high mountains, could be supported for important lengths of time during the Archaean period. By analogy with the present isostatic-topography relationship, a topographic difference of only 3-4 km is the probable maximum for Early Archaean times. The high geothermal gradients and thinner average lithosphere are also likely to have had important effects on the depth of subduction because even the older oceanic lithosphere could not be very thick and so would have little negative buoyancy—probably resulting in two major effects, i.e. only shallow angle subduction and a re-incorporation of oceanic lithosphere into 'normal' mantle at shallower depths than where the phase transitions now become important. (If, therefore, the driving forces are primarily due to density increases in the descending oceanic lithosphere, then it seems unlikely that such forces could have operated in the earliest Precambrian.) Naturally, such broad generalisations must be re-assessed against a more authentic background, indicated by the nature and apparent behaviour of Archaean and Proterozoic continental crustal rocks still preserved today.

The absence of pre-Archaean rocks is, in itself, peculiar as other terrestrial planets, such as Mars, Venus and the Moon, all have crusts still evident at their surfaces, with ages only a little younger than the date of formation of the planets themselves. It seems likely that the heat generation during the gravitational segregation of the Earth's core, combined with radiogenic heat production, resulted in an initially rapid mantle convection state during the first million years or so of the Earth's history. This convection would then become ordered following the completion of the formation of the core. Knowing that running water was present at the time of formation of the oldest preserved continental rocks, it seems likely that for most of pre-Archaean times, and certainly for later times, the temperature of the surface of the Earth was not drastically different from that today, i.e. between the freezing and boiling points of water, but probably close to some 20°C as a world average.

Chapter 1

At such temperatures the rocks at the surface of the Earth must have been solid and so a lithosphere must have existed. It has been proposed (Tarling, 1978a) that this lithosphere was so thin in pre-Archaean times that the oceanic crust which differentiated at the mid-oceanic ridges was dehydrated at very shallow depths during subduction and that secondary 'island arc' differentiation could thus not occur during this period and so continental crust could not be formed. The onset of Archaean times, around 4 000 million years ago, therefore marked the time when the average oceanic lithosphere was sufficiently thick that the older, thicker parts of it were preserved as a cool, undehydrated physical entity down to depths of some 80-100 km. On reaching such depths, based on modern island-arc chemistry, calc-alkaline volcanics would then be produced and the earliest forms of continental crust began to accumulate at the Earth surface. The earliest forms would be sodic, such as those formed today at shallow Benioff Zone depths. The potassium content would increase during the Archaean as subduction proceeded to greater and greater depths reflecting the net increase in the thickness of the lithosphere. The topography during the Archaean would be subdued, probably less than 3-4 km at the start of the Archaean, hence explaining the absence of thick sedimentary accumulations during this period.

The change from Archaean to Proterozoic tectonic regimes occurred between 2 700 and 2 500 million years ago, and was accompanied by extensive magmatic activity (Hart & Brooks, 1977; O'Nions & Pankhurst, 1978), and a change in the composition of the surface continental rocks (Schwab, 1978; Taylor, 1979). Such a change could arise when the net thickness of continental lithosphere reached pressure conditions where its base was more stable for granulite grade than amphibolite grade metamorphic rocks (Tarling, 1978a, 1980c). The onset of Proterozoic times could thus have been initiated by the dehydration of the lower Archaean lithosphere as anhydrous granulite formed. The ascending water and other volatiles would scavenge out other volatiles and lithophile elements, particularly the radiogenic heat-producing elements, which would then become concentrated in the surface rocks, leaving the lower crust and underlying mantle depleted in water and also in the radiogenic heat-producing elements.

This major geochemical event had profound effects because the dehydration of the Archaean continental lithosphere caused major changes in its physical properties. In particular, the depletion in volatile content meant that its melting temperature was raised by 200°C or more, and therefore the level at which viscous flow became possible became very much deeper, presumably comparable with the present lithospheric thickness beneath the Archaean cratons—250-400 km. As a consequence

Introduction: Plate Tectonics

of this thickening, convection in the upper mantle would become concentrated into the oceanic areas where the thinner lithosphere allowed more effective loss of geothermal heat. This concentration of convection would further thin this lithosphere and it seems conceivable that, during the Early Proterozoic, the thickness of the oceanic lithosphere was comparable with that of Archaean times and island arc differentiation processes would be again impaired. The contrast between the two lithospheres would also be further increased because isostatic compensation would take place at very shallow levels in the oceanic lithosphere, and hence oceanic topography would be subdued. Isostatic compensation in the continental lithosphere would be similar to today, allowing the accumulation of thick sedimentary sequences on continental margins and the creation of major mountain areas.

Sea-floor spreading, on this proposed model, has therefore been a feature of the Earth's ocean basins throughout the history of the Earth, but the splitting of continental lithosphere was strongly inhibited during the Early Proterozoic because of its difference in thickness, compared with the oceanic lithosphere, which would mean that convective motions would always break through the very thin oceanic lithosphere rather than the thicker continental lithosphere. On the whole, therefore, the continental blocks would accumulate into a few large blocks which would move as major structural units in response to mantle motions, as indicated by the palaeomagnetic and palaeoclimatic evidence, but they would not be split by such motions. The separation of continental blocks, with sea-floor spreading developing new oceans between the split blocks, would therefore not occur until the strength of the two lithospheres became comparable towards the end of Proterozoic times. Splitting would then occur along the weaker lines within the continental lithosphere when new convective systems evolved in the underlying mantle rocks.

It must, of course, be emphasised that such a paradigm is, at best, schematic, although it does appear consistent with the characteristics of Precambrian terraines. It does, however, have important implications for economic factors. The fact that the continents were in motion means that they would be changing palaeoclimates in response to changes in palaeolatitude, with modifications due to the extreme size of continental blocks during most of the Proterozoic (although the climate would be locally influenced by the distribution of land and sea). The model also predicts shallow angles of subduction during Archaean and Early Proterozoic times, with restrictions in the likely composition of the igneous rocks, and associated metallogenesis, forming in subduction zones. The changing levels of isostatic compensation would also have major influence on the nature of the tectonic evolution along continental margins, as well as in the formation of sedimentary basins, within the continental blocks themselves. However, further consideration of the relationship between plate tectonics and the formation of economically important accumulations of hydrocarbons, coal and metals will be summarised in the final chapter following the more detailed discussions in the chapters succeeding this introduction.

References

Allis, R.G. (1979) A heat production model for stable continental crust. *Tectonophys.*, **57,** 151-165.
Anderson, D.L. (1967) Phase Changes in the Upper Mantle. *Science,* **157,** 1165-1173.
Anderson, D.L. (1975) Structure of the Mantle. *Revs. Geophys. Space Phys.,* **13**(3), 300-302.
Anderson, J.M. & Schwyzer, R.U. (1977) The Biostratigraphy of the Permian and Triassic. Part 4. *Trans.geol. Soc. S. Afr.,* **80,** 211-234.
Arkani-Hamed, J. (1978) Viscous Coupling in a Spherical Body of Temperature-dependent Viscosity. *Tectonophys.,* **50,** 135-145.
Arzi, A.A. (1978) Critical Phenomena in the Rheology of Partially Melted Rocks. *Tectonophys.,* **44,** 173-184.
Aumento, F., Ade-Hall, J.M. & Keen, M.J. (1975) 1974—The Year of the Mid Atlantic Ridge. *Revs. Geophys. Space Phys.,* **13,** 53-65.
Barazangi, M., Isacks, B.L., Oliver, J., DuBois, J. & Pascal, G. (1973) Descent of Lithosphere beneath New Hebrides, Tonga, Fiji and New Zealand: Evidence for Detached Slabs. *Nature,* **242,** 98-101.
Barker, P.F. (1972) Magnetic Lineations in the Scotia Sea. In *Antarctic Geology and Geophysics,* (Ed. R.J. Adie) Universitetsforlaget, Oslo, pp. 17-26.
Barker, P.F. & Burell, J. (1977) The Opening of Drake Passage. *Marine Geol.,* **25,** 15-34.
Benioff, H. (1949) Seismic evidence for the Fault Origin of Oceanic Deeps. *Bull. Geol. Soc. Amer.,* **60,** 1837-1856.
Bickle, M.J. (1978) Heat loss from the earth: a constraint on Archaean tectonics from the relation between geothermal gradients and the rate of plate production. *Earth Planet. Sci. Letters,* **40,** 301-315.
Birch, F. (1952) Elasticity and constitution of the Earth's interior. *J. Geophys. Res.,* **57,** 227-286.
Blakely, R.J. & Cande, S.C. (1979) Marine magnetic anomalies. *Revs. Geophys. Space Phys.,* **17,** 204-214.
Bodvarsson, G. & Lowell, R.P. (1972) Ocean-floor heat flow and the circulation of Interstitial Waters. *J. Geophys. Res.,* **77,** 4472-4475.
Bott, M.H.P. (1971) *The Interior of the Earth.* Edward Arnold, 316 pp.
Bottinga, Y. (1974) Thermal aspects of sea-floor spreading and the nature of the suboceanic lithosphere. *Tectonophys.,* **21,** 15-18.
Bullard, E. (1954) The flow of heat through the floor of the Atlantic Ocean. *Proc. Roy. Soc. Lond.,* **A222,** 408-429.
Cann, J.R. (1974) A model for oceanic crustal structure developed. *Geophys. J. R. astr. Soc.,* **39,** 169-187.
Cann, J.R. & Moore, D.G. (1978) Oceanic basement probed. *Geotimes,* **23**(7), 17-19.
Cathles, L.M. (1975) *The Viscosity of the Earth's Mantle.* Princeton University Press, Princeton, 386 pp.
Chapman, D.S. & Pollack, H.N. (1975) Global heat flow: A new look. *Earth Planet. Sci. Letters,* **28,** 23-32.
Chapple, W.M. & Tullis, T.E. (1977) Evaluation of the forces that drive the plates. *J. Geophys. Res.,* **82,** 1967-1984.

Introduction: Plate Tectonics

Christensen, N.I. & Salisbury, M.H. (1975) Structure and constitution of the lower oceanic crust. *Revs. Geophys. Space Phys.,* **13**(1), 57-86.
Coats, R.R. (1962) Magma type and crustal structure in the Aleutian Arc. *Geophys. Mem. 6.* Amer. Geophys. Union, pp. 92-109.
Cox, A., Doell, R.R. & Dalrymple, G.B. (1963) Geomagnetic polarity epochs and pleistocene geochronology. *Nature,* **198**, 1049-1051.
Daly, R.A. (1940) *Strength and Structure of the Earth.* Prentice-Hall, Englewood Cliffs, 434 pp.
Davis, E.E. & Lister, C.R.B. (1974) Fundamentals of Ridge Crest topography. *Earth Planet. Sci. Letters,* **21**, 405-413.
Davis, E.E. & Lister, C.R.B. (1977) Heat flow measured over the Juan de Fuca Ridge: Evidence for widespread hydrothermal circulation in a highly heat transportive crust. *J. Geophys. Res.,* **82**, 4845-4860.
Dawson, J.B. (1977) Sub-cratonic crust and upper mantle models based on xenolith sites in kimberlite and nephelinite diatremes. *J. Geol. Soc.,* **134**(2), 173-184.
De Long, S.E., Dewey, J.F. & Fox, P.J. (1979) Topographic and geologic evolution of fracture zones. *J. Geol. Soc. Lond.,* **136**, 303-310.
Dewey, J. & Spall, H. (1975) Pre-Mesozoic Plate Tectonics: How far back in Earth history can the Wilson Cycle be extended? *Geology,* **3**, 422-424.
Dickinson, W.R. (1970) Relation of andesites, granites, and derivative sandstones to arc-trench tectonics. *Revs. Geophys. Space Phys.,* **8**, 813-860.
Dickinson, W.R. (1975) Potash-depth (K-h) relations in continental margin and intra-oceanic magmatic arcs. *Geology,* **3**, 53-56.
Elsasser, W.M. (1969) Convection and stress propagation in the upper mantle. In *Application of Modern Physics to the Earth* (Ed. S.K. Runcorn) Wiley, New York, pp. 223-246.
Engdahl, E.R. & Scholz, C.H. (1977) A double Benioff zone beneath the Central Aleutians: An unbending of the lithosphere. *Geophys. Res. Letters,* **4**, 473-476.
Epp, D. & Suyenaga, W. (1978) Thermal contraction and alteration of the oceanic crust. *Geology,* **6**(12), 726-728.
Ernst, W.G. (1975) *Metamorphism and plate tectonic regimes.* Halsted Press, New York, 440 pp.
Forsyth, D. & Uyeda, S. (1975) On the relative importance of the driving forces of plate motion. *Geophys. J. R. astr. Soc.,* **43**, 163-200.
Fowler, C.M.R. (1978) The mid-Atlantic Ridge: structure at 45°N. *Geophys. J. R. astr. Soc.,* **54**(1), 167-183.
Fowler, C.M.R. & Keen, C.E. (1979) Oceanic crustal structure—mid Atlantic Ridge at 45°N. *Geophys. J. R. astr. Soc.,* **56**, 219-226.
Francis, T.J.G. & Porter, I.T. (1973) Median Valley Seismology: The mid-Atlantic Ridge near 45°N. *Geophys. J. R. astr. Soc.,* **34**, 279-311.
Francis, P.W., Moorbath, S. & Thorpe, R.S. (1977) Strontium isotope data for Recent andesites in Ecuador and North Chile. *Earth Planet. Sci. Letters,* **37**, 197-202.
Frisch, W. (1978) A plate tectonics model of the Eastern Alps. In *'Alps Apennines, Hellenides'* (Eds H. Closs, D. Roeder & K. Schmidt) Schweizerbart, pp. 167-172.
Froidevaux, C., Schubert, G. & Yuen, D.A. (1977) Thermal and mechanical structure of the upper mantle: a comparison between continental and oceanic models. *Tectonophys.,* **37**, 233-246.
Girdler, R.W. & Styles, P. (1978) Sea-floor spreading in the western Gulf of Aden. *Nature,* **271**, 615-617.
Gonz, J.H. & Cleary, J.R. (1976) Variations in the structure of the upper mantle beneath Australia from Rayleigh wave observations. *Geophys. J. R. astr. Soc.,* **44**, 507-516.
Goslin, J., Beuzart, P., Francheteau, J. & Le Pichon, X. (1972) Thickening of the Oceanic Layer in the Pacific Ocean. *Marine Geophys. Res.,* **1**, 418-427.
Green, D.H. & Liebermann, R.C. (1976) Phase equilibria and elastic properties of a pyrolite model for the oceanic upper mantle. *Tectonophys.,* **32**, 61-92.

Chapter 1

Gunn, B.M. (1976) A comparison of modern and Archaean oceanic crust and Island-arc Petrochemistry. In *The Early History of the Earth* (Ed. B. F. Windley) Wiley, London, pp. 389-403.

Hales, A.L. (1969) Gravitational Sliding and Continental Drift. *Earth Planet. Sci. Letters,* **6,** 31-34.

Hall, J. & Simmons, G. (1979) Seismic velocities of Lewisian metamorphic rocks at pressures to 8 kbar : relationship to crustal layering in North Britain. *Geophys. J. R. astr. Soc.,* **58,** 337-347.

Hargraves, R.B. (1976) Precambrian Geologic History. *Science,* **193,** 363-371.

Hart, S.R. & Brooks, C. (1977) The geochemistry and evolution of early Precambrian mantle. *Contrib. Mineral. Petrol.,* **61,** 109-128.

Hart, S.R. & Staudigel, H. (1978) Oceanic crust : Age of hydrothermal alteration. *Geophys. Res. Letters,* **5**(12), 1009-1012.

Hasegawa, A., Umino, N. & Takagi, A. (1978a) Double-planed deep seismic zone and upper-mantle structure in the northeastern Japan arc. *Geophys. J. R. astr. Soc.,* **54**(2), 281-296.

Hasegawa, A., Umino, N. & Takagi, A. (1978b) Double-planed structure of the deep seismic zone in the northeastern Japan arc. *Tectonophys.,* **47,** 43-58.

Hasegawa, A., Umino, N., Takagi, A. & Suzuki, Z. (1979) Double-planed deep seismic zone and anomalous structure in the upper mantle beneath northeastern Honshu (Japan). *Tectonophys.,* **57,** 1-6.

Hoffman, P. (1973) Evolution of an Early Proterozoic Continental Margin : The Coronation Geosyncline and Associated Aulacogens of the Northwestern Canadian shield. *Phil. Trans. Roy. Soc.,* **A273,** 547-581.

Holmes, A. (1927) Some problems of physical geology and the Earth's thermal history. *Geol. Mag.,* **64,** 263-278.

Holmes, A. (1944) *Principles of Physical Geology.* T. Nelson & Sons, London, pp. 532.

Hutchinson, C.S. (1976) Indonesian active volcanic arc : K, Sr, and Rb variation with depth to the Benioff zone. *Geology,* **4,** 407-408.

Irving, E., Emslie, R.F. & Ueno, H. (1974) Upper Proterozoic Paleomagnetic Poles from Laurentia and the History of the Grenville Structural Province. *J. Geophys. Res.,* **79,** 5491-5502.

Isacks, B., Oliver, J. & Sykes, L.R. (1968) Seismology and the New Global Tectonics. *J. Geophys. Res.,* **73,** 5855-5899.

Jakes, P. & Gill, J. (1970) Rare Earth Elements and the Island Arc Tholeiitic Series. *Earth Planet. Sci. Letters,* **9,** 17-28.

Jakes, P. & White, A.J.R. (1972) Major and trace elements abundances in Volcanic Rocks of Orogenic areas. *Geol. Soc. Amer. Bull.,* **83,** 29-40.

Jordan, T.H. (1975) The continental tectosphere. *Revs. Geophys. Space Phys.,* **13**(3), 1-12.

Jordan, T.H. & Frazer, L.N. (1975) Crustal and Upper Mantle Structure from Sp Phases. *J. Geophys. Res.,* **80,** 1504-1518.

Jordan, T.H. & Fyfe, W.S. (1976) Lithosphere-asthenosphere boundary. *Geology,* **4**(12), 770-772.

Kent, P.E. (1976) Major synchronous events in continental shelves. *Tectonophys.,* **36,** 87-91.

Kent, P.E. (1977) The Mesozoic development of a seismic continental margins. *J. Geol. Soc. Lond.,* **134**(1), 1-18.

Kono, Y. & Yoshii, T. (1975) Numerical experiments on the thickening plate model. *J. Phys. Earth,* **23,** 63-75.

Kono, Y. & Amano, M. (1978) Thickening model of the continental lithosphere. *Geophys. J. R. astr. Soc.,* **54**(2), 405-416.

Kremp, G.O.W. (1977) The positions and climatic changes of Pangaea and five Southeast Asian Plates during Permian and Triassic times. *Paleo. Data Banks,* **7**(1), 1-21.

Kuno, H. (1959) Origin of Cenozoic petrographic provinces of Japan and surrounding areas. *Bull. Volc.,* **20,** 37-76.

Introduction: Plate Tectonics

Kuno, H. (1966) *Lateral variation of basalt magma across Continental Margins and Island Arcs.* Geol. Surv. Canada, Paper 66-15, pp. 317-335.

Kuznir, N.J. & Bott, M.H.P. (1976) A thermal study of the formation of oceanic crust. *Geophys. J. R. astr. Soc.,* **47**(1), 83-95.

Lambert, R. St. J. (1976) Archaean thermal regimes, crustal and upper mantle temperatures, and a progressive evolutionary model for the Earth. In *The Early History of the Earth* (Ed. B. F. Windley) Wiley, London, pp. 363-373.

Lambert, St.J., Chamberlain, V.E. & Holland, J.G. (1976) The geochemistry of Archean rocks. In *The Early History of the Earth.* (Ed. B.F. Windley) Wiley, London, pp. 377-403.

Langseth, M.G., Anderson, R.N. & Hobart, M.A. (1977) Seafloor heat flow measurements and the extent of water circulation in oceanic crust. *Abs. with Prog. Geol. Soc. Amer.,* **9**(7), 1065.

Le Pichon, X. (1968) Sea-floor spreading and continental drift. *J. Geophys. Res.,* **73**, 3661-3697.

Le Pichon, X., Francheteau, J. & Bonnin, J. (1973) *Plate Tectonics.* Elsevier, 300 pp.

Lilwall, R.C., Francis, T.J.G. & Porter, I.T. (1978) Ocean bottom seismograph observations on the mid-Atlantic ridge near 45°N—further results. *Geophys. J. R. astr. Soc.,* **55**, 255-262.

Lister, C.R.B. (1974) On the penetration of water into hot rock. *Geophys. J. R. astr. Soc.,* **39**, 465-509.

Liu, L-G. (1975) Post oxide phases of olivine and pyroxene and mineralogy of the mantle. *Nature,* **258**, 510-512.

Macdonald, K.C. & Atwater, T.M. (1978) Evolution of rifted ocean ridges. *Earth Planet. Sci. Letters,* **39**, 319-327.

McKenzie, D.P. & Morgan, W.J. (1969) Evolution of triple junctions. *Nature,* **224**, 125-133.

Marsh, B.D. & Kantha, L.H. (1978) On the heat and mass transfer from an ascending magma. *Earth Planet. Sci. Letters,* **39**, 435-443.

McDougall, I. & Tarling, D.H. (1963) Dating of polarity zones in the Hawaiian Islands. *Nature,* **200**, 54-56.

McElhinny, M.W., Taylor, S.R. & Stevenson, D.J. (1978) Limits to the expansion of Earth, Moon, Mars and Mercury and to changes in the gravitational constant. *Nature,* **271**, 316-321.

Meyer, H.O.A. (1979) Kimberlites and the Mantle. *Revs. Geophys. Space. Phys.,* **17**, 776-788.

Mitchell, A.H.G. (1973) Metallogenic belts and angle of dip of subduction zones. *Nature Phys. Sci.,* **245**, 49-52.

Mitchell, A.H.G. (1974) Flysch-ophiolite successions : polarity indicators in arc and collision-type orogens. *Nature,* **248**, 747-749.

Miyashiro, A. (1961) Evolution of metamorphic belts. *J. Petrol.,* **2**, 277-311.

Miyashiro, A. (1972) Metamorphism and related magmatism in plate tectonics. *Amer. J. Sci.,* **272**, 629-656.

Miyashiro, A. (1975) Origin of the Troodos and other ophiolites : a reply to Hynes. *Earth Planet. Sci. Letters,* **25**, 217-222.

Molnar, P. & Atwater, T. (1978) Interarc spreading and Cordilleran tectonics as alternates related to the age of subducted oceanic lithosphere. *Earth Planet. Sci. Letters,* **41**, 330-340.

Molnar, P., Freedman, D. & Shih, J.S.F. (1979) Lengths of intermediate and deep seismic zones and temperatures in downgoing slabs of lithosphere. *Geophys. J. R. astr. Soc.,* **56**, 41-54.

Moorbath, S. (1976) Age and Isotope Constraints for the Evolution of Archaean Crust. In *The Early History of the Earth* (Ed. B. F. Windley) Wiley, London, pp. 351-360.

Chapter 1

Moorbath, S., Thorpe, R.S. & Gibson, I.L. (1978) Strontium isotope evidence for petrogenesis of Mexican andesites. *Nature,* **271,** 437-439.
Moore, J.G., Fleming, H.S. & Phillips, J.D. (1974) Preliminary Model for Extrusion and Rifting at the Axis of the Mid-Atlantic Ridge, 36° 48′ North. *Geology,* **2**(9), 437-440.
Morgan, W.J. (1971) Convection plumes in the lower mantle. *Nature,* **230,** 42-43.
Mori, T. & Green, D.H. (1976) Subsolidus equilibria between pyroxenes in the CaO-MgO-SiO$_2$ system at high pressures and temperature. *Amer. Mineral.,* **61,** 616-625.
Morley, L.W. (1963) Unpublished manuscripts quoted in *Plate Tectonics and Geomagnetic Reversals* by A. Cox, W.H. Freeman and Co., San Francisco, pp. 224-225.
Mutch, T.A. (1979) Planetary surfaces. *Revs. Geophys. Space Phys.,* **17,** 1694-1722.
Newman, P.H. & Falconer, R.K.H. (1978) Evidence for movement between Greenland and Canada along Nares Strait. *Abs. with Prog. Geol. Soc. Amer.,* **10**(7), 463.
Nisbet, E.G. & Fowler, C.M.R. (1978) The Mid-Atlantic Ridge at 37 and 45°N: some geophysical and petrological constraints. *Geophys. J. R. astr. Soc.,* **54,** 631-660.
O'Nions, R.K. & Pankhurst, R.J. (1978) Early Archaean rocks and geochemical evolution of the Earth's crust. *Earth Planet. Sci. Letters,* **38,** 211-236.
Osmaston, M.F. (1977) Some fundamental aspects of plate tectonics bearing on hydrocarbon location. In *Developments in Petroleum Geology* (Ed. G.D. Hobson) Vol. 1, Appl. Sci. Publ., London, pp. 1-52.
Parker, R.L. & Oldenburg, D.W. (1973) Thermal Model of Ocean Ridges. *Nature Phys. Sci.,* **242,** 137-139.
Pollack, H.N. & Chapman, D.S. (1977) On the regional variation of heat flow, geotherms, and lithospheric thickness. *Tectonophys.,* **38,** 279-296.
Poupinet, G. (1979) On the relation between P-wave travel time residuals and the age of continental plates. *Earth Planet. Sci. Letters,* **43,** 149-161.
Press, F. (1971) The Earth and the Moon. *Q. J. R. astr. Soc.,* **12,** 232-243.
Ramberg, I.B., Gray, D.F. & Reynolds, R.G.H. (1977) Tectonic evolution of the FAMOUS area of the Mid-Atlantic Ridge, lat. 35° 50′ to 37° 20′N. *Bull. Geol. Soc. Amer.,* **88,** 609-620.
Reyment, R.A. (1969) Ammonite biostratigraphy, continental drift and oscillatory transgressions. *Nature,* **224,** 137-140.
Richardson, R.M., Solomon, R.C. & Sleep, N.H. (1979) Tectonic stress on the plates. *Revs. Geophys. Space Phys.,* **17,** 981-1019.
Ridd, M.F. (1971) South-east Asia as a Part of Gondwanaland. *Nature,* **234,** 531-533.
Ringwood, A.E. (1958) Constitution of the Mantle (3); Consequences of the olivine-spinel transition. *Geochem. Cosmochim. Acta,* **15,** 195-212.
Ringwood, A.E. (1974) The petrological evolution of island arc systems. *J. Geol. Soc. Lond.,* **130,** 183-204.
Ringwood, A.E. (1975) *Composition and Petrology of the Earth's Mantle.* McGraw Hill, New York, 618 pp.
Ringwood, A.E. (1976) Phase transformations in descending plates and implications for mantle dynamics. *Tectonophys.,* **32,** 129-143.
Ross, C.A. (1979) Late Paleozoic collision of North and South America. *Geology,* **7,** 41-44.
Runcorn, S.K. (1972) Dynamical Processes in the Deeper Mantle. *Tectonophys.,* **13,** 623-637.
Sacks, I.S., Linde, A.T., Rodriquez, B.A. & Snoke, J.A. (1978) Shallow seismicity in subduction zones. *Geophys. Res. Letters,* **5**(11), 901-903.
Sakuyama, M. (1977) Lateral variation of phenocryst assemblages in volcanic rocks of the Japanese islands. *Nature,* **269,** 134.
Sammis, C.G. (1976) The effects of polymorphic phase boundaries on vertical and horizontal motions in the Earth's mantle. *Tectonophys.,* **35,** 169-182.
Sammis, C.G., Smith, J.C., Schubert, G. & Yuen, D. (1977) Viscosity-Depth profile of the Earth's Mantle: Effects of Polymorphic phase transitions. *J. Geophys. Res.,* **82**(26), 3747-61.

Introduction: Plate Tectonics

Schilt, S., Oliver, J., Long, G., Krumhansl, P., Kaufman, S., Jensen, L., Cook, F., Brown, L. & Brewer, J. (1978) Fine structure of the continental lithosphere : Results from the Cocorp deep seismic reflection program. *Abs. with Prog., Geol. Soc., Amer.,* **10**(7), 486.

Schmidt, K. (1978) Review of Earlier Geodynamic concepts. In *Alps Appennines Helleuides* (Eds, Closs *et al.*). Schweizerbart, Stuttgart, pp. 5-9.

Schubert, G. & Turcotte, D.L. (1971) Phase Changes and Mantle Convection. *J. Geophys. Res.,* **76**, 1424-1432.

Schubert, G., Yuen, D.A., Froidevaux, C., Fleitout, L. & Souriau, M. (1978) Mantle Circulation with Partial Shallow Return Flow : Effects on Stresses in Oceanic plates and Topography of the Sea Floor. *J. Geophys. Res.,* **83**, 745-758.

Schwab, F.L. (1978) Secular trends in the composition of sedimentary rock assemblages—Archean through Phanerozoic time. *Geology,* **6**, 532-536.

Shaw, D.M. (1976) Development of the Early Continental Crust. Part 2 : Prearchaean, Protoarchaean and Later Eras. In *The Early History of the Earth* (Ed. B.F. Windley). Wiley, London, pp. 33-53.

Sillitoe, R.H. (1972) A Plate tectonic model for the origin of Porphyry Copper Deposits. *Econ. Geol.,* **67**, 184-197.

Sleep, N.H. (1975) Stress and Flow beneath Island Arcs. *Geophys. J. R. astr. Soc.,* **42**(3), 827-857.

Smithson, S.B. (1978) Modeling continental crust : structural and chemical constraints. *Geophys. Res. Letters,* **5**(9), 749-752.

Smithson, S.B., Shive, P.N. & Brown, S.K. (1977) Seismic reflections from Precambrian crust. *Earth Planet. Sci. Letters,* **37**, 333-338.

Solomon, S.C. (1976) Geophysical restraints on radial and lateral temperature variations in the upper mantle. *Amer. Mineral.,* **61**, 788-803.

Srivistava, S.P. (1978) Evolution of the Labrador Sea and its bearing on the early evolution of the North Atlantic. *Geophys. J. R. astr. Soc.,* **52**, 313-357.

Sutton, J. & Watson, J.V. (1974) Tectonic evolution of continents in early Proterozoic times. *Nature,* **247**, 433-435.

Talwani, M. & Eldholm, O. (1977) Evolution of the Norwegian-Greenland Sea. *Bull. Geol. Soc. Amer.,* **88**, 969-999.

Tarling, D.H. (1973) Metallic Ore Deposits and Continental Drift. *Nature,* **243**, 193-196.

Tarling, D.H. (1978a) *Evolution of the Earth's crust.* Academic Press, London & New York, 446 pp.

Tarling, D.H. (1978b) Geological-Geophysical Aspects of Ice Ages. In *Climatic Change* (Ed. J. Gribbin). Cambridge University Press, Cambridge, pp. 3-24.

Tarling, D.H. (1979) Palaeomagnetic Reconstructions and the Variscan Orogeny. *Proc. Ussher Soc.,* **4**, 233-261.

Tarling, D.H. (1980a) The Geologic Evolution of South America during the Last 200 million years. In *Evolutionary Biology of the New World Monkeys and Continental Drift* (Eds R. L. Ciochon & A. B. Chiarelli). Plenum, New York (in press).

Tarling, D.H. (1980b) Upper Palaeozoic Continental Distributions based on Palaeomagnetic Studies. In *The Terrestrial Environment and the Origin of Land Vertebrates* (Ed. A.L. Panchen). Academic Press, London, pp. 11-37.

Tarling, D.H. (1980c) Lithosphere evolution and changing tectonic regimes. *Quat. J. Geol. Soc., London,* **137**, 459-466.

Tarling, D.H. & Tarling, M.P. (1975) *Continental Drift* (2nd Edition). Doubleday & Co., New York, 140 pp.

Tarney, J. & Windley, B.F. (1977) Chemistry, thermal gradients and evolution of the lower continental crust. *J. Geol. Soc.,* **134**(2), 153-172.

Taylor, S.R. (1979) Chemical composition and evolution of the continental crust : the rare earth element evidence. In *The Earth : Its Origin, Structure and Evolution* (Ed. M.W. McElhinny). Academic Press, London, pp. 353-376.

Chapter 1

Taylor, S.R. & McLennan, S.M. (1979) Discussion in 'Chemistry, thermal gradients and the evolution of the lower continental crust' by J. Tarney & B.F. Windley. *J. Geol. Soc. Lond.*, **136,** 497-500.

Tozer, D.C. (1977) The thermal state and evolution of the Earth and terrestrial planets. *Sci. Prog. Oxf.*, **64,** 1-28.

Veizer, J. (1974) Probable models of the evolution of continental crust and their control by seawater $^{87}Sr/^{86}Sr$ ratios. *Abs., Europ. Geophys. Soc.*

Veizer, J. (1976) $^{87}Sr/^{86}Sr$ Evolution of Seawater during geologic history and its significance as an Index of Crustal Evolution. In *The Early History of the Earth* (Ed. B. F. Windley). Wiley, London, pp. 569-578.

Veizer, J. & Compston, W. (1976) $^{87}Sr/^{86}Sr$ in Precambrian carbonates as an index of crustal evolution. *Geochem. Cosmochim. Acta*, **40,** 905-914.

Vetter, U.R. & Meissner, R.O. (1977) Creep in geodynamic processes. *Tectonophys.*, **42,** 37-54.

Vine, F.J. & Matthews, D.H. (1963) Magnetic anomalies over Oceanic Ridges. *Nature*, **199,** 947-949.

Weissel, K. (1980) Magnetic lineations in the marginal basins of the west Pacific. *Phil. Trans. Roy. Soc., Lond.*, (in press).

White, R.S. (1979) Oceanic upper crustal structure from variable angle seismic reflection-refraction profiles. *Geophys. J. R. astr. Soc.*, **57,** 683-726.

Whitmarsh, R.B. (1975) Axial Intrusion Zone beneath the Median Valley of the Mid-Atlantic Ridge at 37°N detected by Explosion Seismology. *Geophys. J. R. astr. Soc.*, **42,** 189-215.

Windley, B.F. (Editor) (1976) *The Early History of the Earth.* Wiley, London, 619 pp.

Wilson, J.T. (1965) A New Class of Faults and their bearing on Continental Drift. *Nature*, **207,** 343-347.

Yoshii, T. (1975) Regionality of group velocities of Rayleigh Waves in the Pacific and thickening of the plate. *Earth Planet. Sci. Letters*, **25,** 305-312.

Yoshii, T. (1979) A detailed cross-section of the deep seismic zone beneath northeastern Honshu, Japan. *Tectonophys.*, **55,** 349-360.

Chapter 2

Petroleum: Introduction and the Formation and Migration of Hydrocarbons

R. STONELEY & R.J. BAILEY

2.1 General introduction*

It has been said that plate tectonics has not found a single barrel of oil, and two distinguished scientists, Bullard (1975) and Moody (1975), both saw little impact of plate tectonics on the discovery of oil. There is indeed some truth in these views, and opponents of plate tectonics (e.g. Meyerhoff & Meyerhoff, 1974) are still successful in exploration. However, it is not the whole story. What plate tectonics has done is to provide petroleum geologists with a framework within which to review the habitat of oil and gas occurrences. It has provided a basis from which to study the relative distributions of source and reservoir rocks, as expressed through the evolution of palaeogeographies, and to trace the structural development of petroleum-bearing basins.

During the past decade our understanding of the generation, migration and entrapment of petroleum (the term is taken to include both *oil* and natural *gas*) has increased enormously. At the same time the stage has been reached where, over much of the world, there are very few obvious and easily accessible prospects remaining to be drilled. The search now is therefore for more obscure, and generally smaller, accumulations. This will be the case increasingly in the future but an enhanced understanding, combined with vastly improved techniques, provides confidence that this challenge can be faced.

The way ahead is through an integration of all possible lines of geological and related approach embracing many different fields of expertise. Plate tectonics and its implications not only have a fundamental place in this scene but in some cases may provide a unifying factor. At the very least, in building up the integrated picture of the development of a petroleum-bearing basin, plate tectonics will provide guide-lines or constraints to our thinking. However, there are also instances where certain phenomena seem to be explicable only in a plate tectonic context. Once this is recognised in a particular case then there may be grounds for a complete re-interpretation of all data, and this could lead to the discovery of further petroleum accumulations. An example is provided by

*Sections marked thus are written by R. Stoneley.

Chapter 2

the Prudhoe Bay Field, which lies close to the northern coast of Alaska. The main reservoir is a Permo-Triassic sandstone of alluvial origin (Morgridge & Smith, 1972; Jones & Speers, 1976) (Fig. 2.1): the sandstone

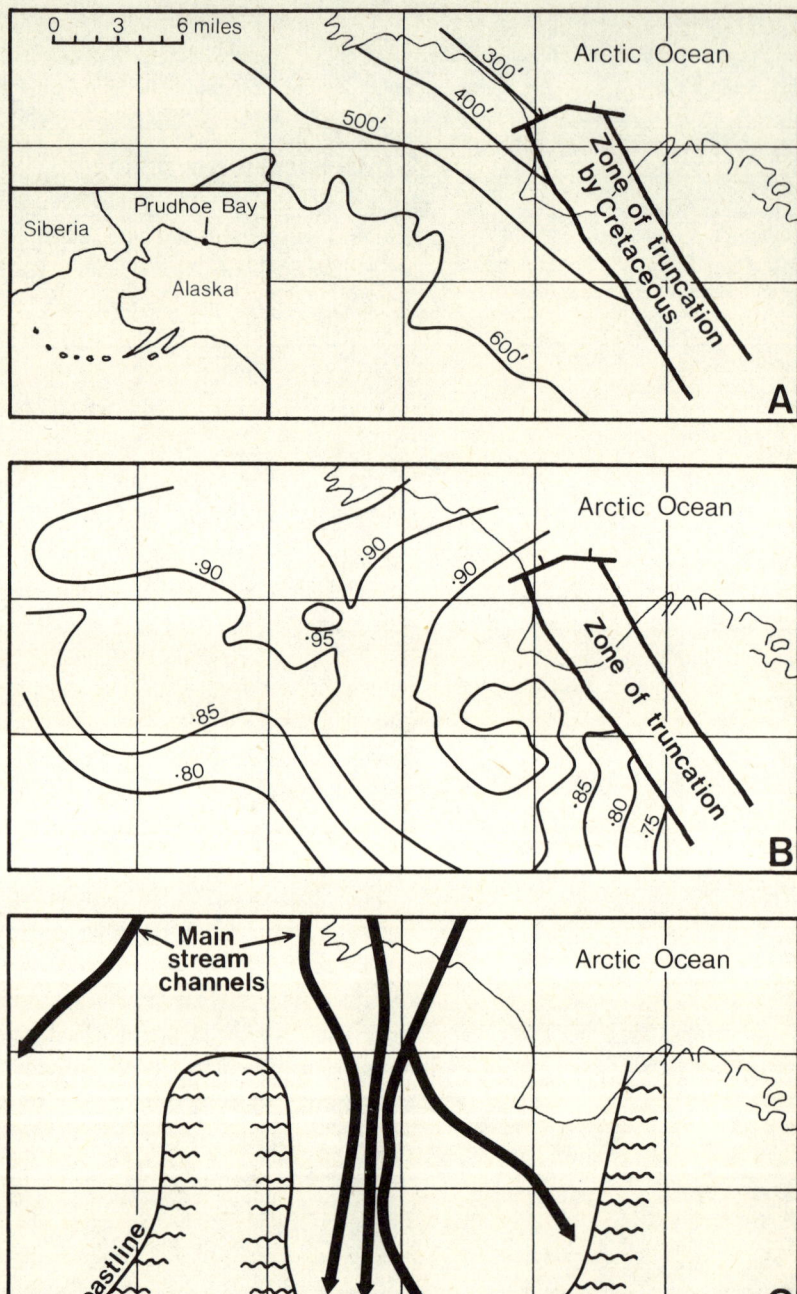

Fig. 2.1. The Prudhoe Bay Field, northern Alaska. The Permo-Triassic Ivishak Sandstone is the principal reservoir of the field and it was derived from a source area on the site of the present Arctic Ocean. (a) Isopachs of the Ivishak Sandstone; (b) sand/shale ratio; (c) idealised palaeogeography of part of the formation (Member B). (After Jones & Speers, 1976.)

Formation and Migration of Hydrocarbons

was not derived from any known onshore region of Alaska, and it thickens generally towards the Arctic coast, close beyond which is a deep oceanic basin. Plate tectonics allow us to suppose that a continental source area for the sand once lay to the north and has drifted away (Rickwood, 1970); it is thus possible to make more confident predictions about the sub-surface distribution and properties of the reservoir sand.

Plate tectonics is the expression of global processes, manifested through sea-floor spreading, subduction, transform-faulting and continental drift. These first-order effects influence the distribution of petroleum initially through the formation and evolution of sedimentary basins, and through their palaeogeographies. Indeed, it can be argued that once these have been established then the more precise controls on petroleum habitat are already inherent in the system and can no longer be attributed directly to plate tectonics. It is this discrepancy in the magnitude of events that has led to scepticism as to the impact of plate tectonics on the study of petroleum. It thus becomes almost impossible and meaningless to try to separate those effects that could be attributed directly to plate tectonics, from those which are an expression of basinal setting or palaeogeography.

In these three chapters the ways in which the concept of plate tectonics have influenced thinking on the habitat of petroleum will be reviewed—thinking which is focussed through basinal setting and palaeogeography. Each of the five basic controls, or prerequisites, for an exploitable petroleum accumulation to exist are considered: *source rock* from which hydrocarbons could be generated, and *heat* to effect that generation are considered in this chapter. Then the development of sedimentary basins in their plate tectonic context is reviewed, followed by a more general consideration of *reservoir rock* to contain free oil or gas, *cap-rock* to contain them within the reservoir, and a *trap* to enable their accumulation in quantity. For the benefit of readers for whom this may be a first contact with the geology of petroleum brief descriptions of some of the general principles involved are included. Chapter 4 then reviews the deep seas in the context of petroleum geology, followed by an overall conclusion concerning the relevance of plate tectonics to petroleum in general.

First, however, let us be clear about what this substance is that is being considered. Petroleum is, after water, the second most abundant fluid in the Earth's crust, and it is a complex and variable blend of several hundred hydrocarbon and subordinate non-hydrocarbon organic compounds. The most prominent constituents are the members of the straight-chain hydrocarbon series (normal paraffins or alkanes), of which the first four, methane, ethane, propane and butane, are gaseous under atmospheric conditions: in strongly decreasing respective abundance, they are the components of natural gas. Generally somewhat more abundant are the naphthenes (cyclo-paraffins or cyclo-alkanes) with one or more saturated carbon rings in the molecule. The branched-chain paraffins (iso-alkanes), and the aromatics (arenes) containing one or more benzene rings in the molecule, comprise lower proportions of crude oils. The non-hydrocarbon constituents include sulphur, oxygen and nitrogen compounds together with minor metallic constituents.

Many lines of chemical and physical evidence combine to prove beyond doubt that most, if not all, of these substances have been derived from organic matter, initially through bacterial activity and subsequently under the influence of geothermal heat. Any contributions from inorganic sources can at best be of extremely minute proportions. Petroleum is formed from the plant and low-order animal remains preserved within the source rock, whence it is expelled into a system of channels, generally between the grains of a porous and permeable rock; it then migrates, either to escape and be lost, or to a trap where further migration is barred. Some aspects of these subjects will be touched on later in this review, but it is not possible to discuss them fully here. The reader is referred to the recent volumes of Tissot & Welte (1978) and Hunt (1979), both of which also give full lists of references.

2.2 Environmental factors in the deposition of petroleum source facies†

It is not immediately evident that a relationship should exist between plate tectonics and petroleum source rocks. Yet if we accept the concensus that such rocks are generally of argillaceous marine facies and contain organic matter deriving, to a significant degree, from microscopic, planktonic algae (Tissot & Welte, 1978), their distribution in time and space is, to some extent, conditioned by lithosphere plate movement. Currently, some 99% of the organic matter in the marine environment ultimately derives from the planktonic algae. These organisms employ solar energy to enforce the combination of CO_2 and H_2O and thus form new organic matter by photosynthesis. This constitutes the primary productivity of the oceans; the shear bulk of the resultant biomass would appear to favour the deposition of organically enriched sediments. However, the by-product of photosynthesis is O_2, and on the death of an alga the solar energy it stored in the form of living tissue is released by vigorous oxidation, often involving aerobic bacteria or grazing zooplankton (Demaison & Moore, 1980). Thus a simple carbon cycle is completed (Fig. 2.2) from which little organic matter, in fact, escapes. In the search for petroleum rocks, therefore, we are looking for ancient environments in which some

† Sections marked thus are written by R.J. Bailey and published by permission of The British Petroleum Company Limited.

Formation and Migration of Hydrocarbons

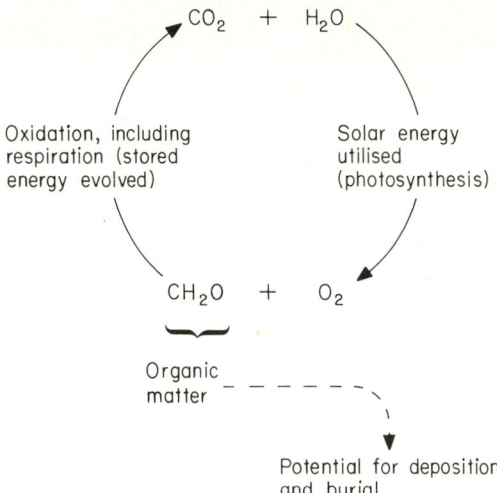

Fig. 2.2. A simplified marine carbon cycle.

fraction of the primary organic productivity escaped this cycle and became a buried store of energy.

Accepting the premise that unicellular algae, phytoplankton, represent the vital ingredient of the detrital organic mix which characterises petroleum source rocks, it is evident that the formation of such rocks is dependent on the co-existence of conditions tending to maximise phytoplankton production—the 'primary productivity' of marine biologists—and conditions tending to minimise the normal oxidative breakdown of the resultant organic matter. Of those factors favouring high primary productivity we can single out the provision of dissolved CO_2 (Fig. 2.2), solar radiation—more or less of the visible spectrum—and vital nutrients such as nitrate, phosphate and trace elements.

The solar energy requirement confines primary productivity to sunlit surface waters, and the *euphotic* zone, thus defined, rarely extends to depths greater than 120 m (Fig. 2.3A). The nutrient requirement is such that high levels of primary productivity can only be maintained in the presence of an undiminished supply of nitrate, phosphate, etc. Normally this is achieved within the euphotic zone by continuous oxidative recycling of the plant tissue formed there (Fig. 2.2); thus the nutrients are passed from generation to generation of phytoplankton. But the deposition of petroleum source rocks demands that some fraction of the primary productivity should escape this cycle, sink to the sea bed and be buried. This represents a drain on the available nutrients which can only be made good by run-off from adjacent lands or by recharge involving the upwelling of deeper, nutrient-enriched waters into the euphotic zone (see below and Fig. 2.4). If neither of these processes occurs surface waters can support only a small standing crop of phytoplankton

35

Chapter 2

and an aquatic desert, such as the Sargasso Sea, is the result (Strickland, 1965; Gordon, 1977).

A further condition for high primary productivity is, therefore, a natural turnover of the water column which ensures that nutrients and CO_2 released by breakdown of organic tissue within and *below* the euphotic zone are well dispersed in the surface waters. It is achieved by vertical turbulent mixing (Fig. 2.3A), resulting from surface wind-stress. This can be effective to depths of 100-200 m in open waters (Bowden, 1975) and may increase its vertical range over open shelves where waves, tides and wind-

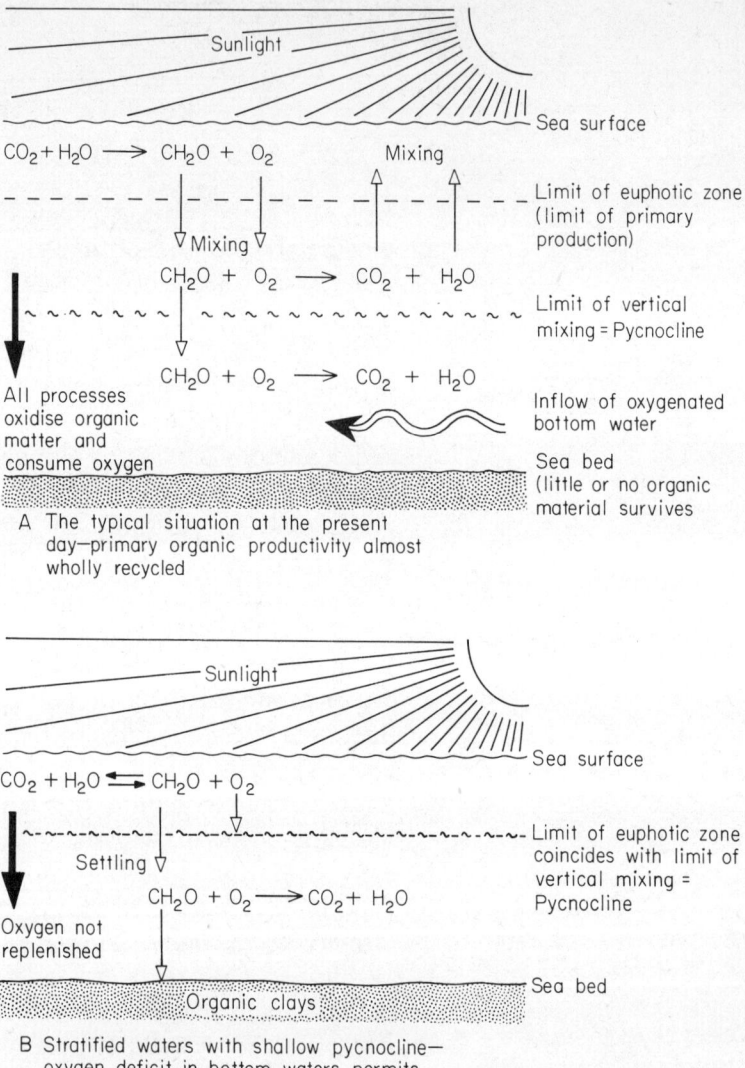

Fig. 2.3. The limit of turbulent vertical mixing and the euphotic zone, showing the way in which their relative depth below sea level can influence the deposition of petroleum source facies.

driven currents interact with the sea bed. Upwelling is a further such effect, involving bodily up-flow of deep, nutrient-rich waters and is most easily understood as a marginal response to the seaward surface water drift induced by prevailing offshore winds (Fig. 2.4).

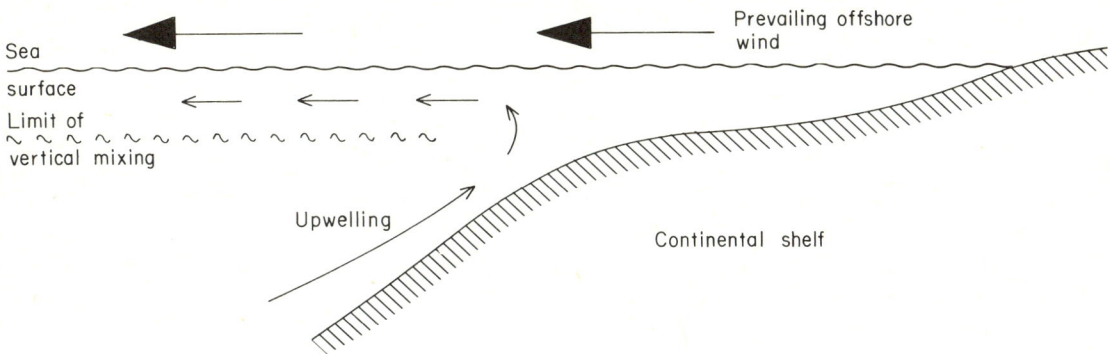

Fig. 2.4. A simple case of marginal upwelling in part of a basin subject to prevailing offshore winds.

The conditions which promote high nutrient levels—well-mixed, shallow waters close to land (Strickland, 1965, Fogg, 1975)—are, in fact, those which typically ensure high levels of dissolved O_2 within and below the euphotic zone. As a consequence little, if any, of the organic matter representing the correspondingly high primary productivity will actually survive to the point of burial. In the North Sea, for example, though primary productivity is exceptionally high, the vigour of the vertical mixing ensures high levels of oxygenation of the waters below the euphotic zone and minimal accumulation of 'algal' organic matter—not least because much of its oxidation is accomplished as 'respiration' by benthic grazing and filter-feeding organisms (Demaison & Moore, 1980).

The way out of the natural cycle of production and oxidation of this kind of organic matter lies in the development of an imbalance in the oxygen budget of the waters *below* the euphotic zone. In these waters, oxygen *demand* is wholly concerned in the breakdown of organic detritus from the euphotic zone, whereas oxygen *supply* is limited by the vigour of vertical mixing. If demand exceeds supply the waters below the euphotic zone become anoxic, and thereafter the rate of breakdown of the detrital organic input is drastically reduced (Demaison & Moore, 1980). This seems the likeliest route to the deposition of petroleum source facies and, in essence, entails the restriction of vertical mixing to the productive euphotic zone.

The equation between wind-stress and the effective depth of vertical mixing implies that a diminution in the former—such as may have characterised the relatively more equable climates of the later Mesozoic (Hallam, 1975; Donn & Shaw, 1977)—would have a bearing on the deposition of petroleum source facies. The absence of climatic extremes may also have deprived the Mesozoic hydrosphere of the dense, cold underflows of polar water which presently ventilate the ocean basins (Fischer & Arthur, 1977). Deposition of petroleum source facies in deep oceanic environments would be correspondingly more likely. This is, of course, merely another aspect of the general process of vertical mixing; and this process is most effectively limited—on all scales from pond to ocean basin—by the development of a stable density stratification of the water column. If, for example, the oxygenated waters of the euphotic zone are less dense than those beneath, buoyancy forces will limit the mixing of the two water masses, downward transport of O_2 will be inhibited and the required imbalance of oxygen supply and demand may develop below the euphotic zone. The result is the ideal situation wherein a productive surface water layer rains organic detritus into oxygen-depleted bottom waters.

In nature significant density stratification typically involves water layers of uniform or gradually varying density separated by pronounced density gradients termed *pycnoclines*. The most common species of pycnocline, a *thermocline*, comes into being where turbulent mixing carries sun-warmed surface water down to some critical depth at which the waning turbulent energy is insufficient to overcome the relative buoyancy of such water. At this depth a thermal gradient stabilises, since further solar energy input only serves to maintain or increase the temperature/density difference between surface and deep-water layers. Because a thermocline is a barrier to downward transport of surface waters, oxygen replenishment in the deep water layer becomes dependent on the slow process of molecular diffusion. Organic detritus from the euphotic zone invades the deeper waters at a much greater rate, and the resultant imbalance in the oxygen budget can give rise to anoxia below the thermocline (Degens & Stoffers, 1976). The ideal condition, wherein the thermocline stabilises within or at the base of the euphotic zone (Fig. 2.3B), is merely a matter of the appropriate degree of surface wind-stress, but it has no *long-term* stability (see below).

Land-locked or semi-enclosed marine basins of moderate to small dimensions (depth less than 2000 m; diameter less than 100 km) may be particularly prone to evolve towards stable thermal stratification, since they are commonly non-tidal and their dimensions control the 'fetch' and therefore the wavelength of the surface wave system, thus moderating the effects of wind-induced vertical mixing. What is more, the nutrient loss occasioned by the deposition and burial of a portion of their primary productivity is likely to be made good by terrestrial run-off.

Marine basins can also achieve the ideal two-layer state outlined above by virtue of salinity differences. Coincidence between the resultant *halocline* and the normal thermocline, in fact, gives rise to the most pronounced density stratification. Haloclines, however, reflect the dynamics of semi-enclosed basins arising either where out-flow of lower salinity surface waters, reflecting a high terrestrial run-off, is coupled with deep in-flow of, say, more normally saline marine waters (for example, the Baltic and the anoxic Black Sea—Grasshoff, 1975), or where surface water in-flow of 'normal' (marine) salinity is needed to counteract evaporative water loss and gravitational sinking of the resultant brines at the nether end of the basin (for example, the Red Sea—Grasshoff, 1975). There may even be instances in which the circulation is partly external to the basin, for example where ground-water movement through buried evaporites generates hypersaline submarine springs that 'flood' the deeper basin (Degens & Paluska, 1979).

Finally, in this review of conditions favouring the deposition of petroleum source facies it is necessary to emphasise an essential feature of the postulated long-term stratification of water masses that is exemplified by modern pycnoclines. The prerequisite is *not* for a completely static condition of stratification. Were this to occur ionic and molecular diffusion of heat and chemical species, possibly coupled with geothermal heating of bottom waters, would, in the geological time frame, blur or even destroy the stratification. What is required in the case of stable halocline development is a circulation—surface flow and deep counter-flow—capable of maintaining the two-layer density stratification, but insufficiently vigorous to maintain the oxygenation of the bottom waters. [At present the Black Sea is in this state, whereas the bottom waters of the Red Sea—outside the 'hot brine' pools—are oxygenated (Grasshoff, 1975).] The requirement, in other words, is for sufficient dynamism in the system to inhibit any tendency to uniformity.

Thermoclines are similarly dynamically dependent, in that pronounced seasonal and (in larger basins) latitudinal changes in surface water temperature and the degree of vertical mixing are necessary to the establishment of the pronounced temperature gradients. They thus tend to be cyclical in their development and would, with a non-seasonal climate, be exceedingly rare phenomena with a life-span limited by the ultimate convergence in temperature of a deeper water layer, warmed by molecular diffusion and geothermal conduction of heat, and a sun-warmed surface layer subject to evaporation, and thus limited in the temperature it can achieve. Possibly, it is the rarity and longevity of Mesozoic thermocline development which is recorded in the broad cyclicity of certain of the classic petroleum source facies such as the Kimmeridge Clay of the British

Isles (Tyson *et al.*, 1979; Gallois, 1976) and of the ubiquitous Cretaceous 'black shale' developments (Ryan & Cita, 1977; Weissert *et al.*, 1979).

What seems clear is that in the development of a 'circulatory' halocline and its (seasonal?) enhancement by the formation of a coincident thermocline, there arise the conditions *most* likely to result in long-term, deep-water anoxia. What is more consideration of simple two-dimensional models suggest that these conditions might conceivably arise even in a completely land-locked basin (Fig. 2.5, see also Brongersma-Sanders, 1971).

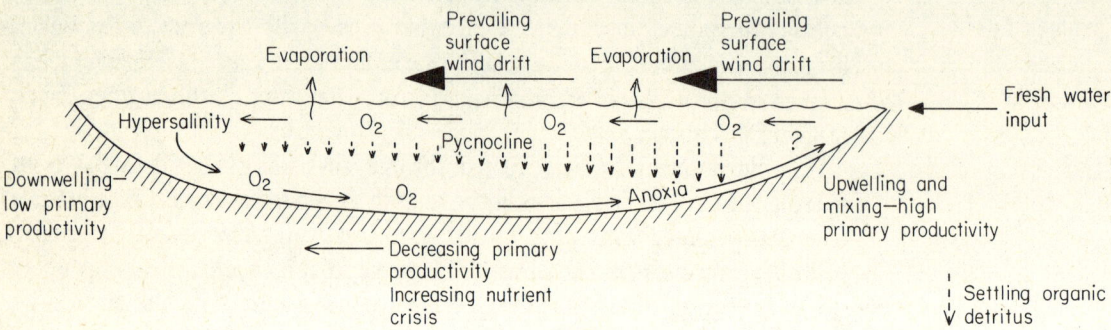

Fig. 2.5. A model to illustrate stable pycnocline development under conditions of thermohaline circulation. Note how the combination of upwelling—giving high surface productivities—and bottom-water oxygen deficit can combine to give a stable anoxic regime in a basin with continuous overturn of the waters.

THE GENERAL BASINAL MODEL

The foregoing analysis suggests that the optimum conditions for the deposition and burial of the fine-grained, phytoplanktonic, organic detritus, thought to be parental to liquid petroleum, are as follows:

1. an upper, mixed, oxygenated, sunlit water layer (100 m or less in depth) where large phytoplankton crops are maintained by a plentiful inflow of nutrients
2. a lower, near-stagnant, anoxic, nutrient-enriched water mass receiving and preserving some of the rain of organic detritus which escapes the upper layer.

It is clear that this association of dissimilar aquatic environments is most likely to be achieved as a consequence of the development of a pycnocline which, in fact, becomes their interface. It has also been suggested that the shallower, semi-enclosed or landlocked, epicontinental, marine or limnic basins are more likely to achieve this condition than the oceans. Open-shelf environments bordering major oceans, on the other hand, are probably those least likely to accumulate petroleum source facies, for the interaction between the oceanic currents, tidal waves and wind waves and the relatively shallow shelf sea bed ensures vigorous mixing and adequate oxygenation of the whole water column. Combine this condition with the resultant proliferation of benthic grazing and filter-feeding organisms and it is difficult to see how organic matter of the requisite type could survive or, in fact, be deposited. Only if the outer shelves were in the past significantly deeper than at present, and thus able to participate in the rare ocean-wide phases of stratification and shallow pycnocline development, e.g. in the Cretaceous (Schlanger & Jenkyns, 1976; Ryan & Cita, 1977), would the deposition of source facies be expected. It must be stressed, however, that this argument relates only to open shelves where there is no impediment to wave and current action. Sheltered areas landward of a major barrier reef system, for example, could well develop stratified waters.

Mention has been made of the influence of the globally warmer and more equable Mesozoic climates in promoting conditions which favour well-developed stratification of major water masses and the consequent anoxia of their bottom waters. A further factor, widely recognised to be of importance in the deposition of petroleum source facies, is the incidence of global marine transgressions. Thus Fischer & Arthur (1977), Hallam & Bradshaw (1979) and Tissot (1979) point to the coincidence between source facies and episodes of transgression, notably those transgressions relating to the apparent Mid to Late Cretaceous global sea-level rise (see also Vail *et al.*, 1977). Since these record the flooding of the subdued 'basin and swell' tectonics of the stable, cratonic continental interiors and, thus, imply the multiplication of the numbers of those semi-restricted epicontinental marine basins believed to be especially prone to stratification, the correlation with anoxic basinal source facies appears easily explicable. However, the results of the Deep Sea Drilling Project clearly indicate that Cretaceous global sea level rise was coupled with a tendency to stratification and anoxia even in the major ocean basins (Schlanger & Jenkyns, 1976; Ryan & Cita, 1977), and that the Albian and Late Cenomanian-Early Turonian represent intervals in which this tendency was strongest. Perhaps the increase in the planetary water surface changed climate still further in the direction of global uniformity, reducing atmospheric and oceanic circulation to the extent that virtually any major water body could be expected to develop, at least intermittently, the type of density stratification described in Section 3.2. There is the suggestion that global sea-level changes in turn correlate with variations in ocean ridge activity (Hays & Pitman, 1973) so that episodes of widespread source facies deposition may ultimately stem from increased vigour of lithosphere plate motion. Whether or not petroleum source facies are quite so fundamentally dependent on plate tectonics, there seems little doubt that certain geological intervals, such as the Cretaceous, are characterised by the prevalence of such facies and, with reference to other equal intervals, by disproportionate amounts of associated oil (Tissot, 1979). In fact, it is now reasonable to consider if, as a first approximation, we can infer adequate petroleum source potential of an epicontinental marine basin solely from, say, the presence of Albian marine shales. If, as seems feasible, this proves to be a useful rule of thumb in oil exploration, it will serve to underline the reality of the relationship between petroleum source rocks and later global evolution, the most obvious attributes of which are the patterns and rates of Phanerozoic plate movement and the associated global rises and falls in sea level. From this standpoint the Cretaceous episode mentioned above may represent the culmination of only one of a series of cycles of organic carbon fixation (see, for example, Fischer, 1980; Ager, 1980; Funnell, 1980; MacKenzie, 1980; Vogt, 1980).

If, in future, the petroleum geologist carries with him a calendar of 'vintage' and 'non-vintage' epochs of petroleum source facies deposition, it will almost certainly reflect the establishment of the, as yet, somewhat tenuous relationship between plate tectonics and such facies.

2.3 The formation of petroleum†

If the relatively rare combination of circumstances which results in the deposition of an organically enriched mud is followed by the progressive burial of this sediment, processes are initiated which can culminate in the formation of a mobile hydrocarbon phase. Tissot & Welte (1978) examine these processes in detail. Here it is sufficient to say that the first few hundred metres of burial are characterised by *diagenesis* during which the proteins and carbohydrates of the organic sediment are attacked microbially. The resultant insoluble organic residue, termed kerogen, on deeper burial undergoes *catagenesis* in which it is degraded thermally from a macromolecular form to produce hydrocarbons. The catagenic process of *maturation* of a petroleum source rock is detectable as an increase in the rocks' yield of extractable hydrocarbons. A simple chemical analysis will thus determine if, and to what degree, it has occurred. Other evidence of catagenesis comes from spores, pollens and the detritus of higher plants which undergo progressive and irreversible changes in optical properties, detectable by transmitted or reflected light microscopy. The correlation of the chemical and optical data thus allows us to identify the onset of catagenesis—the *oil generation threshold*—wholly in terms of optical indices such as spore coloration and vitrinite reflectance (see Tissot & Welte, 1978, p. 449). Use of such indices on well sections from all parts of the world has suggested that temperature is the most important factor in the maturation of petroleum source rocks, but time is also of significance. Thus, if the depth of burial is taken as a crude index of thermal regime, then at comparable burial depths Palaeozoic sediments tend to be measurably more mature than Mesozoic, and the latter more mature than comparable Cenozoic sediments. Expressing the relationship in another

Formation and Migration of Hydrocarbons

way, with an idealised set of Phanerozoic sedimentary basins formed under comparable geothermal regimes, the maturation depth would be least in the Palaeozoic basins and greatest in those of Neogene age. Such relationships can be formalised in time-temperature-depth-maturity diagrams (e.g. Wright, 1980). However, when there is only one potential source formation within a basinal sequence, as is commonly the case, it is necessary to know if, where and when this formation attained maturity. There are various ways of tackling this problem.

Experience has shown that petroleum is probably generated in insignificant amounts at temperatures below 50°C—even when these are maintained over geologically long periods of time (Fig. 2.6). It has also established that petroleum will begin to break down, or 'crack', if temperatures exceed 150°C for even a short time. The 'temperature window' of oil generation, thus defined, can be represented as a range of depths of burial: either by reference to some global geothermal gradient for sedimentary basins in general, or by reference to a geothermal gradient known or judged to be appropriate to the basin in question. Knowing the age and configuration of the source formation and the thickness of the

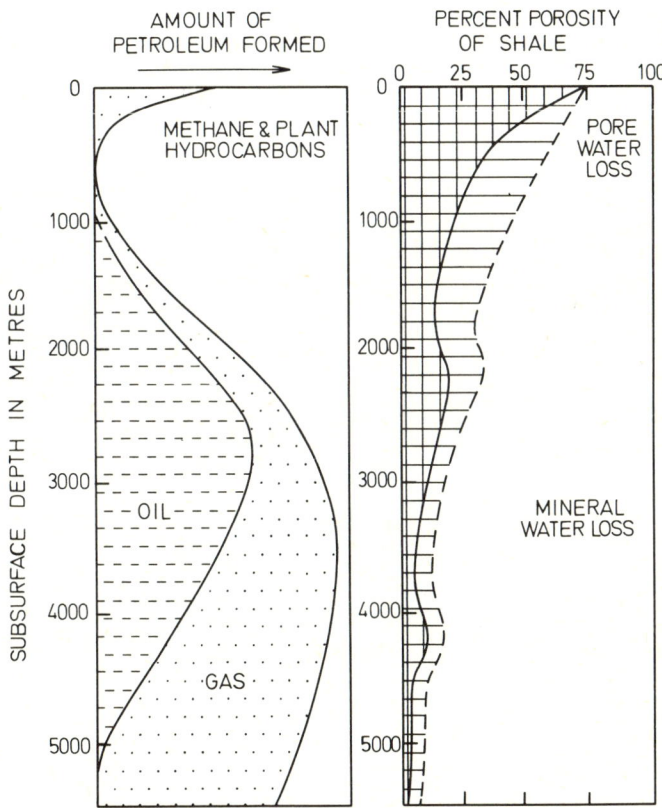

Fig. 2.6. Shale compaction—hydrocarbon generation relationship as a function of burial depth. The effect of time is neglected in this diagram and a geothermal gradient of 3.5°C/100 m is assumed, i.e. an average continental geothermal gradient. The generation of petroleum is generally considered to be mainly dependent on temperature, while the compaction is primarily determined by the weight of overlying sediments, although rapid sedimentation can result in 'overpressures' developing. Normal equilibrium porosity is shown as double hatched, while overpressurised shale porosities are shown as single hatched.

Chapter 2

time units which successively buried it, we can construct burial histories for various parts of the basin and thus establish whether or not the source rocks anywhere entered the temperature window of oil generation. More sophisticated theoretical models of time-temperature-depth-maturity relationships have been proposed, tested and reviewed (e.g. Tissot & Welte, 1978; Wright, 1980). It must be stressed, however, that these either derive from detailed geochemical and optical studies, or are held to be valid theoretical models because they make predictions more or less coincident with the results of such studies. They have a basin-wide rather than a global application, since in the global context catagenesis involves too many variables. This can be illustrated by reference to the 'temperature window', the crudest and most widely applicable of the empirical estimates of maturation depths. This embraces, at its low-temperature end, the 50°C threshold of oil generation thought to characterise old—Palaeozoic—basins and clearly would be inappropriate to Neogene basins where the threshold temperature of oil generation is commonly found to be higher. Couple with this the observable basin-to-basin variability in present and past geothermal gradient and it is not difficult to arrive at a situation in which the 50°C isotherm is both inappropriate as an index of the maturation threshold and, by virtue of the basin's history of anomalously low geothermal gradients, is inaccurately identified in terms of burial depth. The net effect would be to infer the thermal maturity of great volumes of rock which are, in fact, wholly immature.

Clearly, maturity assessments should be conducted on a basin-by-basin basis and should ideally stem from determination of spore coloration, vitrinite reflectance and extractible hydrocarbon ratios of samples of known age and burial history. (Cassou *et al.*, 1977, provide a good example of this approach.) This is certainly one of the chief reasons for the current vogue in the oil industry for collaborative drilling of deep stratigraphic tests in undrilled or little explored basins. Surface heat-flow measurements can also be used to delineate basins, or areas of basins, where high geothermal gradients prevail. However, in the absence of these two kinds of data a good general knowledge of the type and age of the target sedimentary basin enables the petroleum geologist to avoid the more obvious errors exemplified in the previous paragraphs. As we shall see plate tectonic theory is useful in this context and, in the particular instance of Atlantic continental margin basins, may eventually be predictive of the geothermal history of the sedimentary sequence (Chapter 4).

2.4 The migration of petroleum †

Whereas the formation of hydrocarbons is relatively well understood, their movement from the source rock into a reservoir and their movement and entrapment within that reservoir, i.e. the general processes of migration, present a number of puzzles. It is evident, for example, that the highly porous watery mud in which the organic source material originally resides will, by the time the ambient temperatures and pressures approach those of significant hydrocarbon generation, have been reduced to an indurated, largely dewatered and effectively impermeable mudrock. By what means can the hydrocarbons migrate from this rock; and, given their ability to do so, how is it that similar rocks, often at shallower depths, constitute an effective seal (Section 3.4)?

In attempting to answer such questions we can at the outset make a distinction between *primary migration*, that is, the hydrocarbons' movement within and out of the source rock, and *secondary migration* occurring when the hydrocarbons have escaped from the source rock and are moving with more freedom in carrier beds, along fault planes and within reservoir rocks. It is also important to recall at every stage in the following review of these processes that migration of both kinds occurs via minute fissures, pores and capillaries that are saturated by warm, high-pressure and, generally, highly saline waters.

PRIMARY MIGRATION

To describe what must occur in primary migration is easier than to describe how it occurs. What we believe to have migrated is a range of hydrocarbons embracing methane (mol. wt 16) at the one extreme and asphaltenes (mol. wt sometimes in excess of 5000) at the other. The media through which these migrate are compacted argillaceous rocks, arguably impenetrable to oil since they can be shown commonly to behave as semi-permeable membranes with respect to ionic solutions (Neglia, 1979). However, given that the process is a reality, the essential question seems to be whether the hydrocarbons were 'active', having their own migratory impetus, or 'passive', forming a solute or suspension borne through the rock by another fluid phase such as water. (Whichever is the case we are presumably thinking in terms of migration down a pressure gradient.)

One difficulty in accepting the 'passive' interpretation with water as the carrier fluid is that most of the water in the system has been gradually expelled by compaction preceding catagenesis and that which is left is effectively bound to clay-mineral surfaces. The answer appears to be to invoke a change in the pore waters of the source rock which coincides with the thermal maturation of hydrocarbons. Amongst such changes are: a temperature-induced increase in the solvent properties of the pore water with respect to hydrocarbons (Price, 1980); overpressuring of the pore system as a consequence of thermally induced de-watering of montmorillonite-type clay minerals (Powers, 1967; Burst, 1969); and overpressuring of the pore waters as a consequence of their thermal expansion (Barker, 1972). In each instance the water moves and carries the hydrocarbons. In general, however, chemical data tend to weigh against both the molecular solutions and the colloidal suspension of petroleum hydrocarbons that would be required if primary migration depended on water as the carrier phase (Tissot & Welte, 1978). Nor, seemingly, can we invoke aqueous flushing of bubbles or globules of hydrocarbon for the water pressures anticipated would be inadequate to force such entities through the minute inter-pore capillaries and throats in the source rock (Hobson, 1973; Hobson & Tiratsoo, 1975; Tissot & Welte, 1978).

Perhaps another unsatisfactory aspect of the 'passive' mechanisms of primary migration is their dependence on the coincidence of not necessarily related phenomena: the formation of the hydrocarbons and the creation of the aqueous overpressures necessary to expel them from the source rock. Admittedly recent studies (Cassou *et al.*, 1977) have confirmed the coincidence of the de-watering of montmorillonite and the onset of catagenesis, but not all source rocks contain significant amounts of montmorillonite.

It seems more satisfactory, if not actually nearer the truth, to argue that primary migration occurs as a direct consequence of the generation of petroleum—that primary migration and quite possibly the early stages of secondary migration are no more than the extremely slow 'explosion' which occurs when kerogens attain their maturation temperatures. This, in essence, is the view advanced by Hedberg (1979, 1980). He argues that free methane generation during catagenesis creates internal fluid pressures which ultimately may exceed the overburden pressure on the highly impermeable source rock. The result is the development of microfractures which permit the escape of the gas and the flushing and/or gaseous solution transport of the liquid hydrocarbons. Whether the microfractures persist during catagenesis, or are only spasmodically induced, they seem to represent the most likely means to effect primary migration.

Tissot & Welte (1978) take a similar view though without the expressed dependence on methane for the creation of the 'explosive' overpressures. Thus they conclude (1978, pp. 284, 298) that the generation of low molecular weight hydrocarbons from polymeric kerogens causes a great increase in molecular volumes and, in an impermeable rock, a corresponding development of centres of high pressure which may induce microfracturing. They stress, however, the spasmodic nature of the process and the need for it to occur many times if significant primary migration is to take place.

Also, the idea that hydrocarbons provide their own migratory impetus—are self-propelled, in effect—contains the germ of an explanation of why a certain minimum amount of kerogen, by weight, seems to be a prerequisite for the generation of migratory petroleum. The same could be said of the other 'active' mechanism of primary migration which involves the oil flowing along a filamentous kerogen network, much as it would along a wick (McAuliffe, 1978).

SECONDARY MIGRATION

Overpressuring to the degree invoked here, involving microfracturing of the essentially impermeable petroleum source rock, also entails forcible invasion of the pore space of contiguous and more permeable reservoir rocks, or of natural migration pathways such as sand streaks or faults, which will conduct the petroleum to the reservoir. Price (1980), for example, sees the forcible dilation of fault planes consequent upon the development of abnormal fluid pressures as providing the only possible avenues of vertical secondary migration in deep basins.

The invasion of the coarser pore system which characterises a reservoir or carrier bed permits the free hydrocarbon phases to aggregate as bubbles or globules which will migrate down the pressure gradient resulting from their 'explosive' generation. This forced early secondary migration must necessarily give way to migration controlled by pore hydrodynamics, the relative buoyancy of the hydrocarbon phase and the relative sizes of the bubbles/globules and the capillaries connecting the pore system. We can envisage, for example, the hydrocarbons ascending faults and invading levels in the sedimentary sequence where compaction de-watering of argillaceous sediments can provide a hydrodynamic drive, or where the porosity and permeability of the coarser clastic units are such as to involve no impediment to the buoyant segregation of hydrocarbons. The distances of 'up dip' secondary migration occasionally appear to be measured in tens of kilometres, but more commonly it has a pronounced vertical component. Thus Price (1980) argues that numerous of the World's major oil-producing basins are characterised by the predominance of vertical secondary migration which utilise faults as the vital link between deeply buried source rocks and the shallower, porous and permeable sandstone and limestone reservoirs.

Where the reservoir rock configuration constitutes a trap (Section 3.3) and the overlying sediments provide a seal (Section 3.4) it is generally found that the relative buoyancy of the hydrocarbons, with respect to the pore waters, determines the formation of an oil or gas pool. This perhaps underlines a distinction between the early and forceful phase of secondary migration which utilises faults and thin carrier beds to bring the hydrocarbons into juxtaposition with a (usually shallower) reservoir, and the later less forceful secondary migration which controls the accumulation of the hydrocarbons in the freer milieu of this reservoir.

Given the availability of a source rock, it is evident that the accumulation of commercially significant amounts of petroleum is *next* dependent on that rock attaining thermal maturity. If it can do so relatively early in its burial history and/or at relatively shallow burial depths, then so much the better, for the greater the time and burial depth needed to achieve maturity, the greater the likelihood that associated reservoirs, sandstones in particular, will have suffered diagenetic reduction in porosity and permeability (Klemme, 1975). Even supposing that secondary migration via faults obviates any need for juxtaposition of source and reservoir, early and shallow generation still increases the likelihood that migrating hydrocarbons will encounter an undegraded reservoir. On these grounds, and presupposing the presence of source, Neogene basins with high geothermal gradient, or with a combination of high sedimentation rate and 'normal' geothermal gradient, should prove to be particularly prolific in oil. Conversely, such basins with high sedimentation rate and unusually low geothermal gradient must carry higher exploration risks. By virtue of their relative youth, it is comparatively simple to arrive at an understanding of the tectonic milieu of such basins and thus, using plate tectonics as predictor, indicate other areas where similar kinds of basins might occur, whether of Neogene or of greater age.

High sedimentation rate coupled with unusually low geothermal gradient is, for example, often characteristic of the fore-arc region of a zone of plate convergence. The back-arc basins by contrast often have high heat-flow. This could determine exploration strategy in the circum-Pacific belts of plate convergence. Similarly it might be argued that the combination of high sedimentation rate and high/normal geothermal gradient particularly characterises zones of thinning and rifting of continental crust—either simple pull-apart or that which associates with major transcurrent faults. Thus, the Los Angeles Basin, among the World's most prolific having produced close on 6.5 billion barrels from a Neogene section characterised by rift tectonics, rapid sedimentation and high geothermal gradient, might become our yardstick and plate tectonics could probably show us where to find more of the same. The trouble is that in doing so it might lead us straight past another Ghawar oilfield (Thralls, 1955; Thralls & Hasson, 1956)!

References

Ager, D.C. (1980) Major marine cycles in the Mesozoic. Abstr. in *Geol. Soc. Lond. Newsletter*, **9**,(2) 18-23.
Barker, C. (1972) Aquathermal pressuring—role of temperature in development of abnormal pressure zones. *Bull. Amer. Assoc. Petrol. Geol.*, **56**, 2068-2071.

Bowden, K.F. (1975) Oceanic and estuarine mixing processes. In *Chemical Oceanography*, Vol. 1 (Eds J.P. Riley & G. Skirrow) 2nd Edition. Academic Press, London, pp.1-41.

Brongersma-Sanders, M. (1971) Origin of major cyclicity of evaporites and bituminous rocks—an actualistic model. *Marine Geol.*, **11**, 123-144.

Bullard, E.C. (1975) Overview of plate tectonics. In *Petroleum and Global Tectonics* (Eds A.G. Fisher & S. Judson) Princeton University Press, Princeton, pp. 5-19.

Burst, J.F. (1969) Diagenesis of Gulf Coast clayey sediments and its possible relation to petroleum migration. *Bull. Amer. Assoc. Petrol. Geol.*, **53**, 73-93.

Cassou, A-M., Connan, J. & Porthault, B. (1977) Relations between maturation of organic matter and geothermal effect as exemplified in Canadian East Coast offshore wells. *Bull. Canad. Petrol. Geol.*, **25**, 174-194.

Degens, E.T. & Stoffers, P. (1976) Stratified waters as a key to the past. *Nature*, **263**, 22-26.

Degens, E.T. & Paluska, A. (1979) Hypersaline solutions interact with organic detritus to produce oil. *Nature, (Lond)*, **281**, 666-668.

Demaison, G.J. & Moor, G.T. (1980) Anoxic environments and oil source bed genesis. *Organic Geochem*, **2**, 9-31.

Donn, W.L. & Shaw, D.M. (1977) Model of climate evolution based on continental drift and polar wandering. *Bull. Geol. Soc. Amer.*, **88**, 390-396.

Fischer, A.F. (1980) Two Phanerozoic super cycles. Abstr. in *Geol. Soc. Lond. Newsletter*, **9**(2), 18-23.

Fischer, A.G. & Arthur, M.A. (1977) Secular variations in the pelagic realm. In *Deep-water carbonate environments* (Eds H.E. Cook & P. Enos) Soc. Econ. Paleon. Mineral. Special Publ. 25, pp. 19-50.

Fogg, G.E. (1975) Primary productivity. In *Chemical Oceanography*. Vol. 22, no. 5 (Eds J.P. Riley & G. Skirrow) 2nd Edition. Academic Press, pp. 385-453.

Funnell, B.M. (1980) Modulation of autocorrelation by biological processes. Abstr. in *Geol. Soc. Lond. Newsletter*, **9**(2), 18-23.

Gallois, R.W. (1976) Coccolith blooms in the Kimmeridge clay and the origin of the North Sea Oil. *Nature*, **259**, 473-475.

Gordon, D.C. (1977) Variability of particulate organic carbon and nitrogen along the Halifax-Bermuda section. *Deep Sea Res.*, **24**, 257-270.

Grasshoff, K. (1975) The hydrochemistry of landlocked basins and fjords. In *Chemical Oceanography* Vol. 2 (Eds J.P. Riley & G. Skirrow) 2nd Edition. Academic Press, London, pp. 455-597.

Hallam, A. (1975) *Jurassic Environments* Cambridge University Press, Cambridge, 280 pp.

Hallam, A. & Bradshaw, M.J. (1979) Bituminous shales and oolitic ironstones as indicators of transgression and regression. *J. Geol. Soc. Lond.*, **136**, 157-164.

Hays, J.D. & Pitman, W.C. (1973) Lithospheric plate motion, sea level changes and climatic and ecological consequences. *Nature*, **246**, 18-22.

Hedberg, H.D. (1979) Methane generation and petroleum migration. *Oil & Gas J.*, 7/5/79, 186-192.

Hedberg, H.D. (1980) Thoughts on petroleum migration. *U.N. Internat. Meeting Petroleum Geology, Beijing, China.*

Hobson, G.D. (1973) The occurrence and origin of oil and gas. In *Modern Petroleum Technology* (Eds G.D. Hobson & W. Pohle) 4th Edition. Applied Science, London, pp. 1-25.

Hobson, G.D. & Tiratsoo, E.N. (1975) *Introduction to Petroleum Geology* Scientific Press, Beaconsfield.

Hunt, J.M. (1979) *Petroleum Geochemistry and Geology* W.H. Freeman & Co., San Francisco. 617 pp.

Jones, H.P. & Speers, R.G. (1976) Permo-Triassic reservoirs of Prudhoe Bay Field, North Slope, Alaska. In *North American Oil and Gas Fields* (Ed. J. Braunstein) Am. Assoc. Petrol. Geol., Mem. 24, pp. 23-50.

Chapter 2

Klemme, H.D. (1975) Giant oilfields related to their geologic setting—a possible guide to exploration. *Bull. Can. Petrol. Geol.*, **23**, 30-66.

McAuliffe, C.D. (1979) Oil and gas migration—chemical and physical constraints. *Amer. Assoc. Petrol. Geol. Bull.*, **63**, 761-781.

McKenzie, F.T. (1980) Global tectonic control of Phanerozoic long-term cycling of carbon. Abstr. in *Geol. Soc. Lond. Newsletter*, **9**(2), 18-23.

Meyerhoff, A.A. & Meyerhoff, H.A. (1974) Tests of Plate Tectonics. In *Plate Tectonics—Assessments and Reassessments* (Ed. C.F. Kahle) Amer. Assoc. Petrol. Geol., Mem. 23, pp. 43-145.

Moody, J.D. (1975) Distribution and geological characteristics of giant oil fields. In *Petroleum and Global Tectonics* (Eds A.G. Fischer & S. Judson) Princeton University Press, pp. 307-320.

Morgridge, D.L. & Smith, W.B. Jr (1972) Geology and Discovery of Prudhoe Bay Field, Eastern Arctic Slope, Alaska. In *Stratigraphic Oil and Gas Fields - Classification, Exploration Methods and Case Histories* (Ed. R.E. King) Am. Assoc. Petrol. Geol. Mem. 16, pp. 489-501.

Neglia, S. (1979) Migration of fluids in sedimentary basins. *Bull. Amer. Assoc. Petrol. Geol.*, **63**, 573-597.

Powers, M.C. (1967) Fluid release mechanisms in compacting marine mudrocks and their importance in oil exploration. *Bull. Amer. Assoc. Petrol. Geol.*, **51**, 1240-1254.

Price, L.C. (1980) Utilization and documentation of vertical oil migration in deep basins. *J. Petrol. Geol.*, **2**, 353-387.

Rickwood, F.K. (1970) The Prudhoe Bay Field. In *Geological Seminar on the North Slope of Alaska* (Eds W.L. Adkison & M.M. Brose) Am. Assoc. Petrol. Geol. Pacific Sec., pp. L1-L11.

Ryan, W.B.F. & Cita, M. (1977) Ignorance concerning episodes of oceanwide stagnation. *Marine Geol.*, **23**, 197-215.

Schlanger, S.O. & Jenkyns, H.C. (1976) Cretaceous oceanic anoxic events: causes and consequences. *Geol. en Mijn.*, **55**, 179-184.

Strickland, J.D.H. (1965) Production of organic matter in the primary stages of the marine food chain. In *Chemical Oceanography*, Vol. 1 (Eds J.P. Riley & G. Skirrow) 1st Edition. Academic Press, London, pp. 478-610.

Thralls, W.H. (1955) Ghawar oilfield, giant among giants: London, *The Oil Forum*, **9**, 121-123.

Thralls, W.H. & Hasson, R.C. (1956) Geology and oil resources of eastern Saudi Arabia. In *Symposium Sobre Yacimentos de Petroleo y Gas*. XX Int. Geol. Cong., Mexico, II, pp. 9-32.

Tissot, B. (1979) Effects on prolific petroleum source rocks and major coal deposits caused by sea level changes. *Nature*, **277**, 463-465.

Tissot, B.P. & Welte, D.H. (1978) *Petroleum Formation and Occurrence* Springer-Verlag, Berlin, Heidelberg & New York. 538 pp.

Tyson, R.V., Wilson, R.C.L. & Downie, C. (1979) A stratified water column environmental model for the type Kimmeridge Clay. *Nature*, **277**, 377-380.

Vail, P.R., Mitchum, R.M. & Thompson, S. (1977) Seismic stratigraphy and global changes of sea level, Part 4: Global cycles of relative changes of sea level. In *Seismic stratigraphy—applications to hydrocarbon exploration* (Ed. C.E. Payton) Amer. Assoc. Petrol. Geol., Mem. 26, Tulsa, USA.

Vogt, P.R. (1980) Periodicity of global magmatic cycles. Abstr. in *Geol. Soc. Lond. Newsletter*, **9**(2), 18-23.

Weissert, H., McKenzie, J. & Hochuli, P. (1979) Cyclic anomix events in the Early Cretaceous Tethys Ocean. *Geol.*, **7**, 147-151.

Wright, N.J.R. (1980) Time, temperature and organic maturation—the evolution of rank within a sedimentary pile. *J. Petrol. Geol.*, **2**, 411-425.

Chapter 3

Petroleum: the Sedimentary Basin
R. STONELEY

3.1 The classification of basins

A sedimentary basin may be defined, for present purposes, as an element of the Earth's surface where sediments have accumulated to a significant thickness: lateral dimensions are measured generally in the order of hundreds of kilometres and thicknesses in the order of kilometres. Most basins have involved some degree of synsedimentary tectonic subsidence, but others are developed where sediments prograde into pre-existing deep-water areas, for example, beyond the edge of the continental shelf. It is to the sedimentary basins that the search for petroleum turns, and each basin presents its own particular problems. Attempts to classify them are useful, since they provide a basis for organising knowledge and thinking, as well as for comparison. In the estimation of the possible reserves of an unexplored basin, for example, it is logical to compare the geological conditions prevailing there with those of known basins and, with caution, to use this comparison in a preliminary evaluation.

Many classifications of sedimentary basins have been proposed (e.g. Bally, 1975; Klemme, 1975). There is no doubt of the value of these attempts to reach an overall understanding, although the variability of nature is such that if the process is taken too far, then each individual basin would fall into a category of its own. Furthermore, the tectonic setting of a basin may change during the course of its history, so that the deposits of one regime may succeed those of a completely different regime: this is a problem that was recognised many years ago (e.g. Kay, 1951).

For these reasons a very simple and broad classification is employed based on plate tectonic processes. It highlights a problem, however, that is inherent in all applications of plate tectonics to the geological past. Plate tectonics is defined through present events and any interpretation of the past record must involve extrapolation and comparisons that are, to a certain extent, subjective. Thus, we may classify a particular basin by comparison with modern analogues, but would we be correct? The problem is compounded where the basin has been through different plate tectonic environments: for example, the western Canadian basin, extending northwards from Alberta, may be considered as an original inactive continental

Chapter 3

margin, that has been involved in collision with another continental element and subsequently evolved in relation to an active margin. It is, therefore, sometimes necessary to assign a basin to different plate tectonic classifications at different stages of its evolution! This also is a subjective process, but the attempt to classify a basin should be made in terms of the interpreted plate tectonic setting *at the time the beds in question were deposited.*

Basins can be classified according to their association with (Fig. 3.1):
1. crustal spreading
2. crustal destruction (subduction)
3. transform faulting
4. intraplate environments.

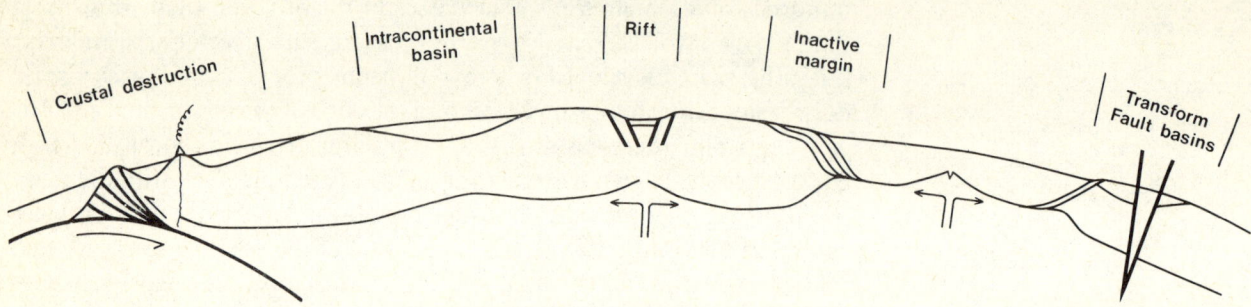

Fig. 3.1. The plate tectonic environments of the primary classes of sedimentary basins.

The last of these, the intraplate basins, evolve and, in some cases, were formed through processes that are still poorly understood and probably not directly attributable to plate tectonics (e.g. Stoneley, 1969; Beaumont, 1978). Each of these categories carries certain geological associations that are of fundamental importance to petroleum. Some generalisations are permissible, although they are certainly not exclusive and, if taken too far, can be misleading. The discussions in this section are necessarily very superficial and references are quoted as a guide to further reading: those that have been selected generally give a list of further references and are from recent, readily accessible sources.

3.2 Basins associated with crustal spreading

The basins in this category fall within the well-known 'rift-drift', or divergent continental margin, sequence of events (e.g. Thompson, 1976) (Fig. 3.2). The concern is essentially with spreading centres that develop within a continental block, since a new one within an oceanic realm is unlikely to lead to the development of a sedimentary basin of interest.

An initial uplift may herald the formation of a new ocean, but the evidence suggests that this is not always the case. Tensional rifting then leads to the emplacement of mantle material within the rift as an incipient

Petroleum: the Sedimentary Basin

STAGE 1– Uplift & fracturing

STAGE 2– Rifting

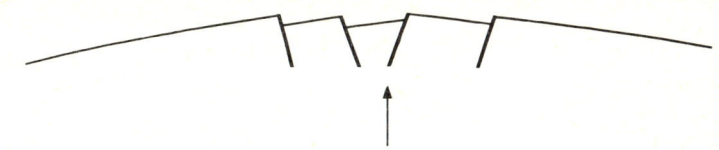

STAGE 3– Further rifting; marine invasion

STAGE 4– Initiation of new ocean

STAGE 5– Ocean formation; development of inactive margins

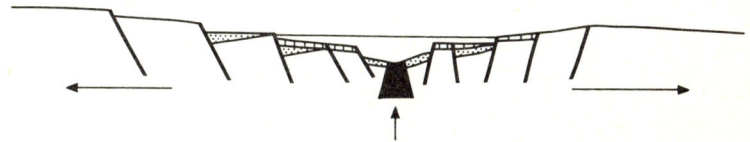

Fig. 3.2. The 'rift-drift' sequence of basin development, from the initial uplift and rifting to the formation of inactive continental margins.

ocean, which subsequently widens until a change in the global distribution of spreading centres takes place. The formation of sedimentary basins accompanies the rifting stage, but the subsequent evolution of the new inactive continental margins thus formed falls within the intraplate category 4.

The classical examples of stages of the 'rift-drift' sequence are the East African rift valleys, the Gulf of Suez, the Red Sea and the Atlantic Ocean. Upwarping is associated with the African rift system (e.g. Baker

Chapter 3

et al., 1972) and the deposits are terrestrial and lacustrine, together with basaltic and alkaline volcanics. Marine invasion is a somewhat later stage, and restricted environments often lead rapidly to the precipitation of evaporites, well exemplified in the Gulf of Suez (e.g. Robson, 1971; Hassan & El-Dashlouty, 1970). Rifting may become complex, as in the southern part of the Red Sea, before the initial emplacement of mantle-derived material in the axial zones (Lowell *et al.*, 1975). Thereafter, the continental margins will continue to separate indefinitely, gradually reaching the stability of inactive shelves, well illustrated by such west African basins as Angola and Gabon (see below, p. 66).

In some cases it seems that the sequence has been interrupted and did not reach the stage of new ocean formation: such a basin is termed an *aulacogen*. The North Sea is generally regarded as an aulacogen that was aborted after the marine invasion but prior to the formation of new oceanic crust (e.g. Ziegler, 1977; Hay, 1978; Pegrum & Mounteney, 1978): it is an instructive example (Fig. 3.3). There is no evidence of initial upwarp: rather a broad depression was filled with Permian desert

Fig. 3.3. North Sea cross-sections, approximately east-west. A. Central North Sea; B. Northern North Sea. Phases of rifting date from the Permian to Mid Cretaceous, following which broad subsidence took place. DP, Devonian and Permian; Z, Zechstein (mainly evaporites); T, Triassic; J, Jurassic; K, Cretaceous; P-E, Palaeocene-Eocene; O-Pl, Oligocene-Pleistocene. (After Ziegler, 1977.)

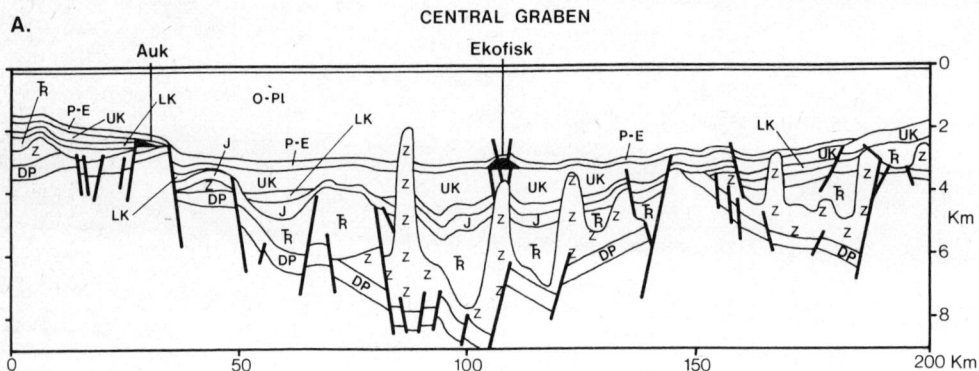

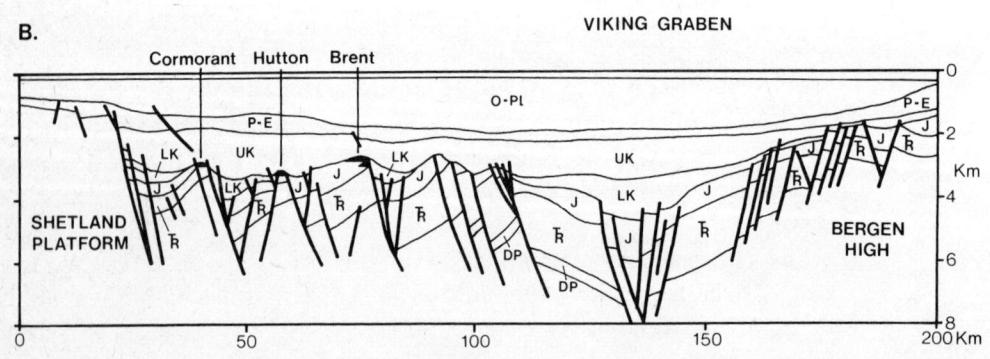

deposits (e.g. Glennie, 1972) and then carbonates and evaporites. Rifting started in the Permian but was intermittent, becoming especially pronounced in the Late Jurassic and Early Cretaceous: after this, however, as spreading was transferred to the North Atlantic, only broad subsidence continued (Kent, 1977). The actual mechanisms involved in the North Sea development are still uncertain (e.g. Ziegler, 1975; Osmaston, 1977).

PETROLEUM

Although minor quantities of gas may be encountered, the early stage of rifting is unlikely to be associated with significant petroleum potential, since the generally narrow troughs contain little or no source rock. However, once marine invasion has taken place a regime is established that is very important. Restricted seas can become enriched in organic matter and heat-flow is high, so that the source and generation potential are good; clastic detritus, and sometimes reefal growths, provide good reservoir rocks; evaporites form cap-rocks and also, perhaps, diapiric structures which, together with tilted fault blocks and drape structures, help to ensure abundant traps. One possible disadvantage may be that the geothermal gradient can be so high that sometimes the zone of gas generation is entered at relatively shallow depth: this is apparently the case in the southern Red Sea (Lowell *et al.*, 1975).

The North Sea basin (Fig. 3.4), extending southwards into northern Germany and adjacent countries, again provides an excellent illustration (e.g. Selley, 1976; Ziegler, 1975; papers in Woodland, 1975). Gas, largely derived from the underlying Carboniferous Coal Measures of the preceding tectonic cycle, is widespread in the south in the Permian basal deposits, both in the Rotliegendes desert sands and in the early marine limestones of the Zechstein: the traps are in block-faulted and warped structures. Oil occurs in these reservoirs in the Auk and Argyll Fields, but it is believed to have been derived from a younger source (Brennand & van Veen, 1975; Pennington, 1975). Jurassic (especially Kimmeridgian) and Lower Cretaceous source rocks have fed penecontemporaneous sandstone reservoirs, of both deltaic and deep-water facies, in tilted blocks produced by intermittent rift-faulting in central and northern parts of the basin: in the south, however, traps in these beds are predominantly caused by salt motions (halokinetic), as also are those such as Ekofisk in which oil has accumulated in the Chalk of the southern Norwegian and Danish sectors. Gentle drape and compaction structures over deeper fault blocks provide the traps for oil in Tertiary reservoirs, as in the Forties Field (Walmsley, 1975).

Chapter 3

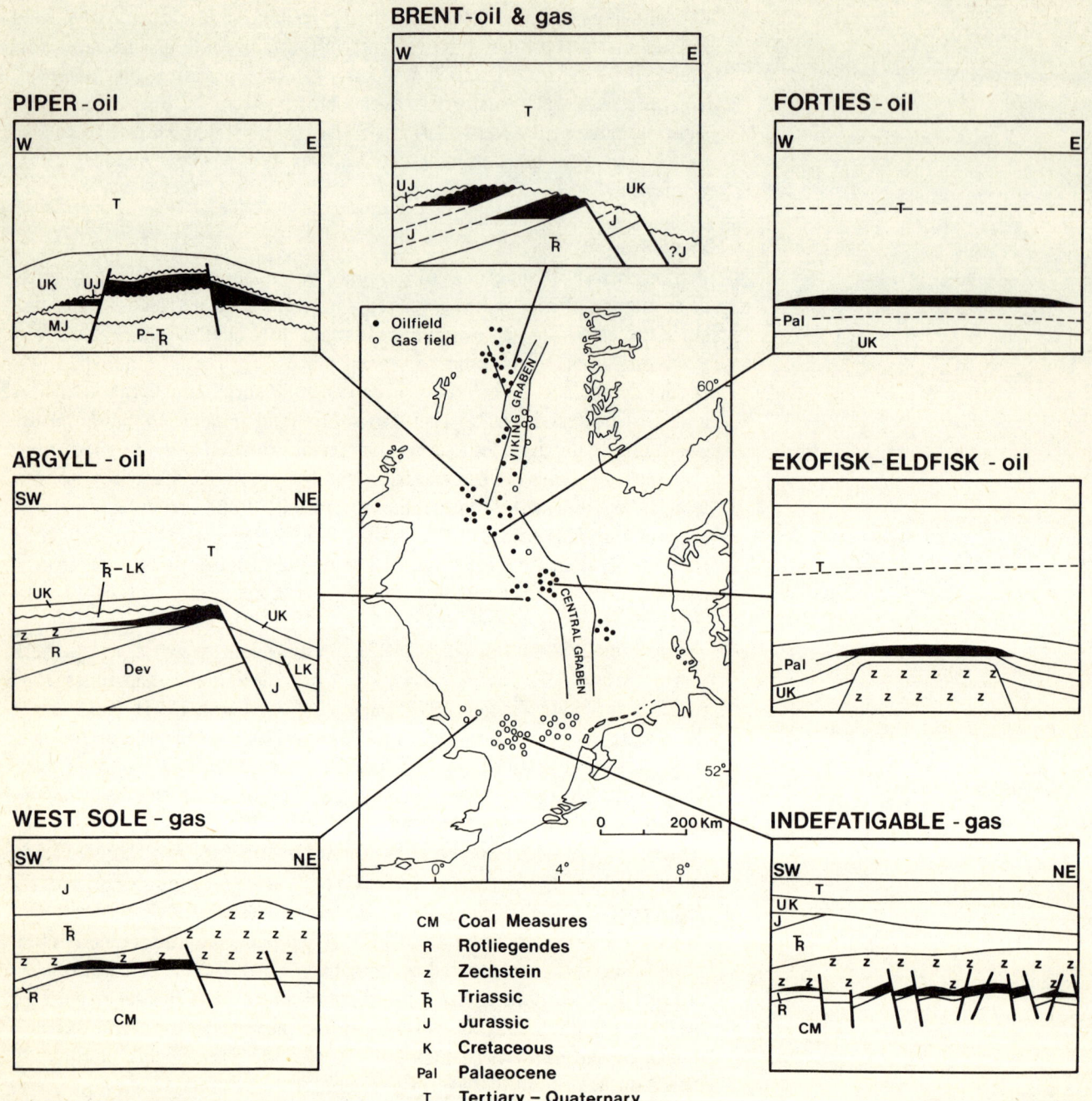

Fig. 3.4. Diagrams showing types of petroleum accumulations in the North Sea. (Based on papers in Woodland, 1975.)

Petroleum: the Sedimentary Basin

3.3 Basins associated with crustal destruction

The basins that are developed in association with belts of plate convergence are very varied. Too close comparisons or generalisations are dangerous, especially since some basins falling under this heading may have evolved from a former inactive continental margin. Thus, for example, the Zagros basin of south-west Iran represented part of an inactive shelf from the Permian into the Neogene, and only then was involved in continental collision and subduction. Because in any classification distinctions have to be made, shelves that are tectonised as a result of collision will be considered in the context of the intraplate basins.

A convenient subdivision is provided by the island-arc environments, as follows: fore-arc/outer-arc, inter-arc basin, volcanic-arc, back-arc basin (Fig. 3.5). These environments can be recognised, with caution, both

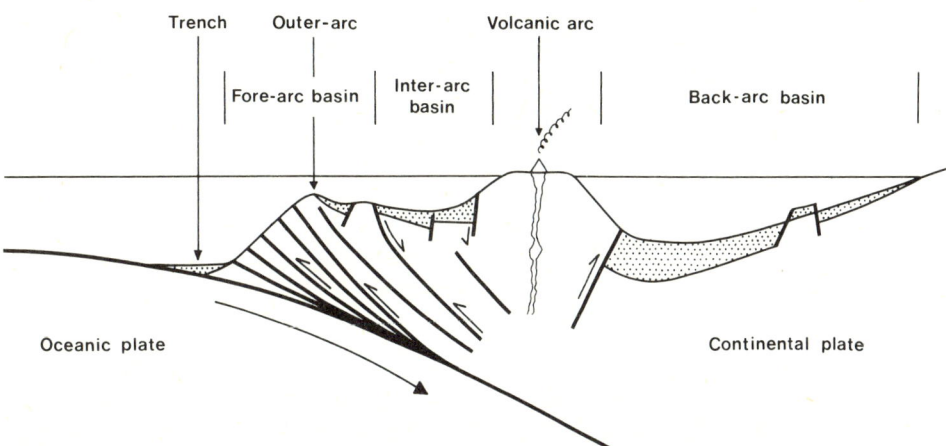

Fig. 3.5. Generalised classification of tectonic elements and basins in a zone of crustal destruction. In an oceanic realm, the back-arc basin will be one-sided and may be expanded through secondary crustal spreading.

in entirely oceanic settings and also where oceanic crust underthrusts a continental margin: indeed, a lateral passage along the strike from the one to the other can sometimes be observed. Thus, for example, the Aleutian Arc passes from an oceanic realm eastwards into continental Alaska. The various basins wil be considered following this geotectonic subdivision.

FORE-ARC BASINS

Fore-arc regions have received a lot of attention during the past few years, and an excellent review has been published recently by Dickinson & Seely (1979). This section deals largely with the deposits of an oceanic trench at the outcrop of a subduction zone. These, together sometimes with slivers of material stripped off the down-going oceanic plate, are piled up above the plane of subduction in a series of imbricate slices on the inner side of the trench. This stack may lead upwards more or less

Chapter 3

directly to the volcanic arc, as seen commonly in the more oceanic cases; or it may lead to the growth of a non-volcanic, outer island arc; or, at an active continental margin, to a series of 'coastal mountain ranges'.

In general, the sediments of such regions are immature, rapidly deposited and thick: when imbricated, the total pile may reach thicknesses of several kilometres. Turbidites, derived both from uplifted earlier sediments and from volcanic arc rocks, predominate and slump deposits are common. Of course, grain flow sands may be present and, if the rate of sedimentation outstrips subsidence, then neritic or even littoral or terrestrial beds can accumulate. Rather short-lived reefs are known. Lateral and vertical facies changes are often rapid, so that reservoir bodies are commonly of limited extent.

Structurally, the fore-arc regions are frequently complex, as a result of the compressional imbrication and also of instability-induced normal faulting. In addition, heat-flow tends to be low, so that any potential source rock is often thermally immature even at considerable depths.

Petroleum

Fore-arc regions are clearly not among the most attractive for petroleum exploration and few commercially viable fields have been discovered. Not only are most geological factors rather unfavourable, but they combine to make exploration difficult; reservoirs of limited extent and complex structure may also be associated with rugged topography, the Gulf of Alaska Cenozoic basin being an extreme example (Fig. 3.6) (Stoneley, 1967). Best known perhaps amongst the oilfields in fore-arc regions are the coastal fields of Ecuador and Peru (Fig. 3.7). They have accumulated in regressive Tertiary sediments of deep to shallow water facies, partly at least in small complexly normal-faulted 'successor' basins also subject to low-angle gliding, above the older rocks of the Coastal Ranges (e.g. Lonsdale, 1978; Travis *et al.*, 1976).

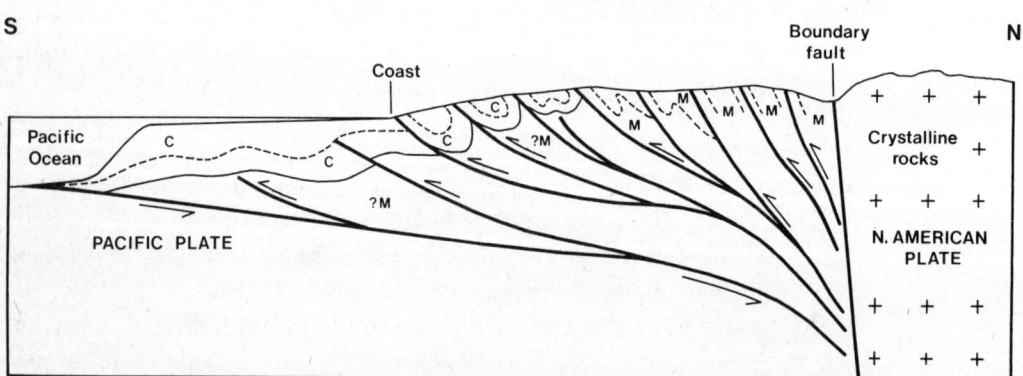

Fig. 3.6. Sketch of the inferred tectonics in the Gulf of Alaska region at an active continental margin. (Based on Stoneley, 1967.)

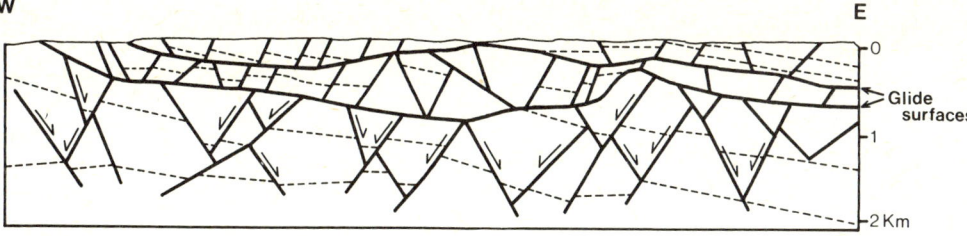

Fig. 3.7. Talara, coastal Peru. Natural scale cross-section through the oilfields area. The block-faulting and glide structures illustrated are in the Eocene and younger paralic deposits of a regressive successor basin, overlying deep-water clays of a fore-arc environment. (After Travis *et al.*, 1976.)

INTER-ARC BASINS

These basins lie between the inner volcanic and the outer non-volcanic arcs: they are often, in the literature, still included under the heading of 'fore-arc basins' (e.g. Seely, 1979). In a regime of continental margin growth as a result of subduction, these basins will generally overlie marginal or trench deposits. They are generally narrow and filled primarily with clastic deposits, derived from both arcs and, sometimes, from the continent itself. Structural styles can be either gently compressive or tensional, depending on the expression of the underlying subduction zone.

Petroleum

The Cook Inlet of southern Alaska (Fig. 3.8), may be cited as one of the few examples of petroliferous inter-arc basins (Kirschner & Lyon, 1973; Fisher & Magoon, 1978). It has partly faulted margins, both normal and strike-slip, and contains more than 9000 m of non-marine Tertiary sediments. Oil and gas are found in the lower part of this Tertiary sequence, in sharp compressive anticlines. There is evidence, however, that the hydrocarbons originated in the underlying Mesozoic beds which, at least in part, appear to have accumulated on a temporarily stable shelf, during a period of interruption of subduction (Fisher & Magoon, 1978).

BACK-ARC BASINS

The third environment of thick sedimentation associated with active plate margins lies behind the volcanic arc. Its development can be very varied, depending on whether it is continental or oceanic, and whether or not it is associated with secondary back-arc crustal spreading. The basins are typically asymmetrical with a steep, often thrust-faulted, outer margin against the volcanic arc and a non-tectonic inner margin which, however, may be affected by basement block-faulting.

The deposits include clastics derived from both margins: from the volcanic arc and, in a continental environment, from the craton behind. The coarser detritus, however, may not reach the centre of the basin where, as for example in Sumatra, argillaceous and carbonate rocks can accumulate

Chapter 3

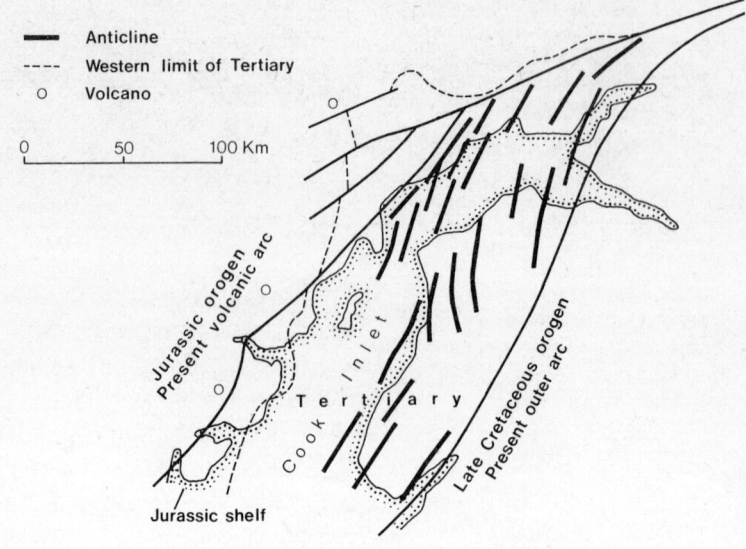

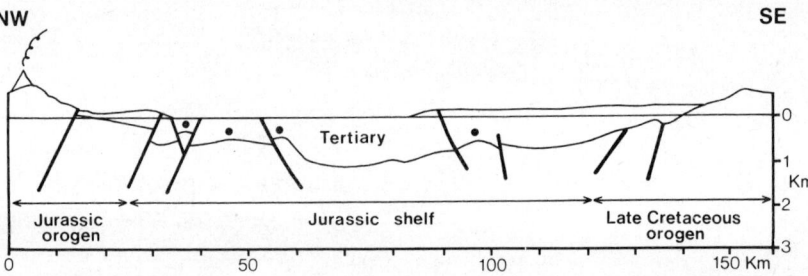

Fig. 3.8. The Cook Inlet Basin, Alaska. A Tertiary inter-arc basin overlying a southwards accreting Mesozoic fore-arc. Oil and gas in the continental Tertiary deposits were derived from a Mesozoic shelf facies belt that has escaped serious orogeny. (Based on Kirschner & Lyon, 1973.)

(de Coster, 1975). In a continental environment the filling of back-arc basins is commonly molassic, derived largely from an orogen along the volcanic belt. The deposits can be coarsely clastic and of either continental or marine facies: the tendency is for them to become generally less marine upwards, in which case petroleum can be expected to be confined to the lower parts of the sequence. This distribution may be emphasised if the basin, and indeed the whole continental margin, evolved from an earlier inactive shelf. Such is the situation in the Oriente basin, lying east of the Andes in Ecuador and Peru (e.g. Feininger, 1975; Zuñiga y Rivero et al., 1976), where Middle Cretaceous and older shelf sediments underlie thick Upper Cretaceous and Cainozoic molassic deposits (Fig. 3.9).

Lateral transitions from one crustal setting to another are known. Thus, for example, the tectonic strike passes from the continental back-arc basin of Burma southwards into oceanic Andaman Sea, where back-arc spreading is taking place (Curray et al., 1979); farther south again, it resumes a continental environment in Sumatra (Fig. 3.10). Although the situation

Petroleum: the Sedimentary Basin

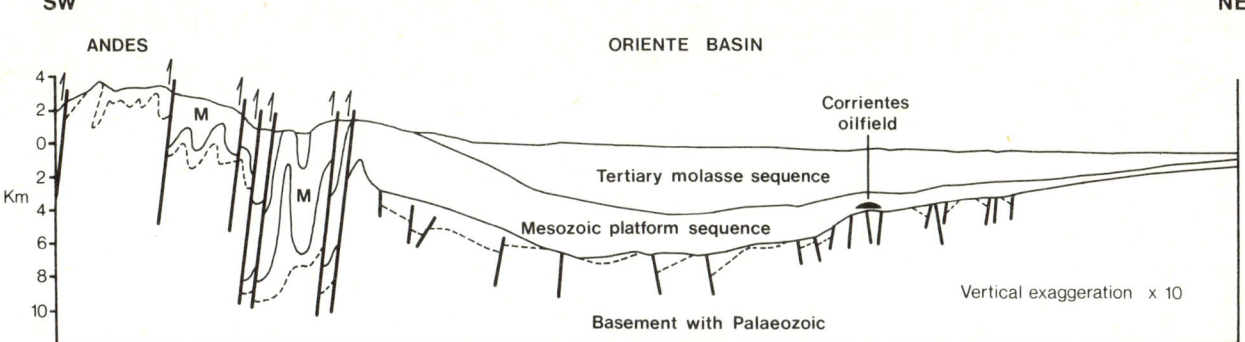

Fig. 3.9. The Oriente Basin of Peru has evolved as a back-arc basin since the Late Cretaceous from an earlier inactive continental margin. (Based on Zuñiga y Rivero et al., 1976.)

is complicated by strike-slip faulting, the development of the volcanic arc is suppressed in the vicinity of the Andaman Sea and nearly all of the sediment is derived from the Asian mainland.

Petroleum

The wide variety in crustal environment and sedimentary filling of back-arc basins produces a corresponding variation in the occurrence of oil and gas. Reservoir rocks could, as a rule, be virtually assured, either as clastics or as carbonates, to the extent even that adequate cap-rock might not always be present. The development of source rocks will mainly be dependent upon the oceanographical environments, but they are especially likely to accumulate in basins of limited areal extent where organic matter may be concentrated. Heat-flow generally is high, particularly in the outer parts of the basin adjacent to the volcanic arc, although it decreases towards the continent: the possibility exists, for example, that the rather heavy oil found in the Oriente of Peru has migrated through considerable distances, up-dip eastwards towards the Brazilian Shield (Feininger, 1975). Structural traps, once away from the basin margin adjacent to the volcanic arc and sometimes associated orogen, are predominantly related to block movements in the underlying continental basement. The examples of Sumatra, Burma and the basins east of the Andes have already been referred to: they are indeed the most important of the back-arc producing regions.

3.4 Basins associated with transform faulting

Where there is transpression across a major strike-slip fault, or where there are deviations from its course, then the resultant stresses can lead to differential crustal warping. This is most likely to happen where transform faults pass through continental regions due to the presence of crustal inhomogeneities. Small basins are believed to have been formed in this manner, and they are commonly subject to second-order folding,

Chapter 3

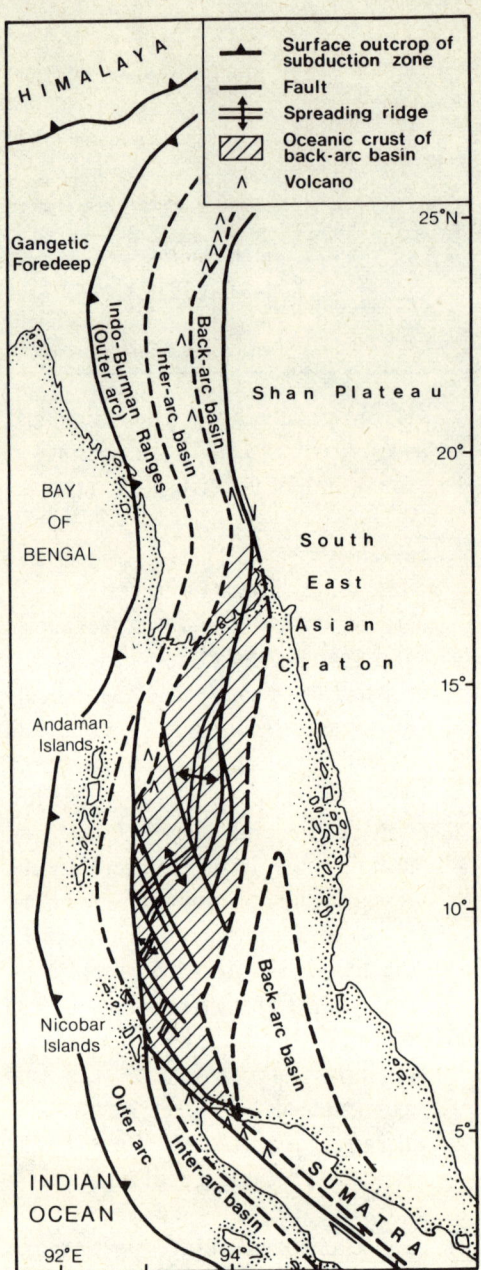

Fig. 3.10. Lateral crustal transitions of a back-arc basin, from continental in Burma, through oceanic in the Andaman Sea (oblique back-arc spreading) to continental in Sumatra. (After Curray et al., 1979.)

thrusting and normal faulting. They are developed, to a greater or lesser extent, in association with most of the world's great strike-slip faults within continental blocks. Such basins may be formed, filled with sediment and uplifted very rapidly. The sediments can range over short

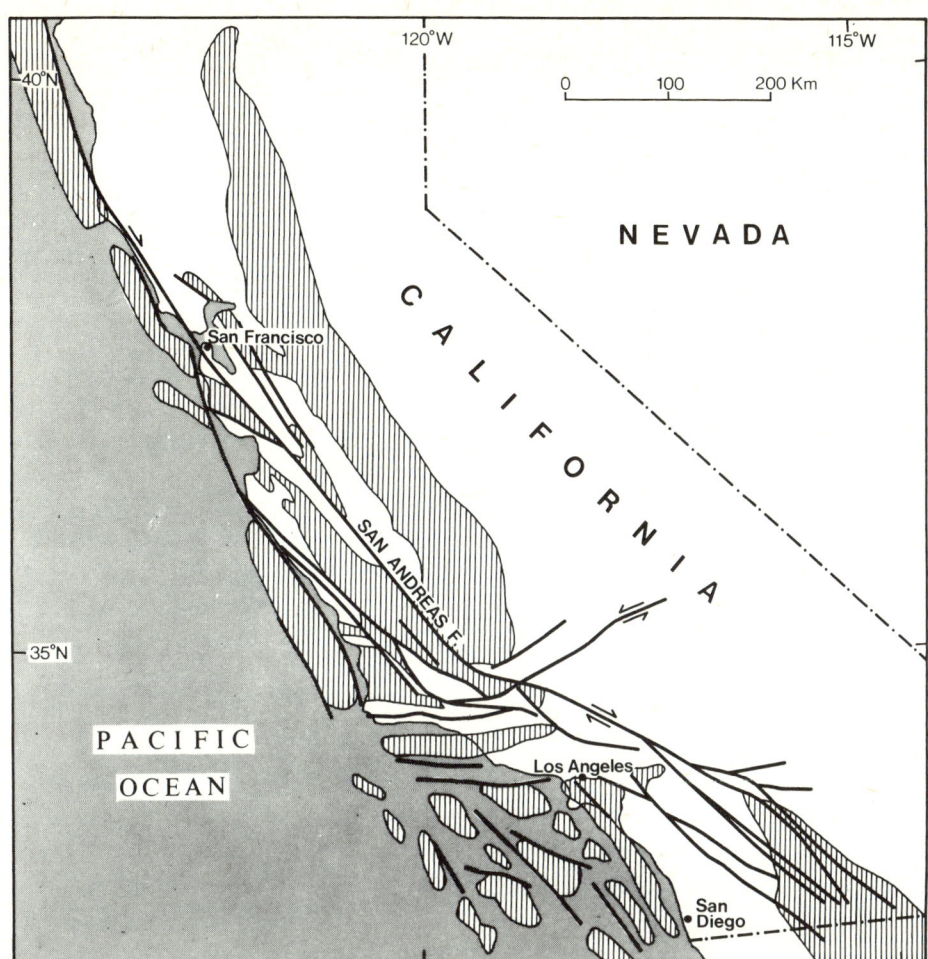

Fig. 3.11. Generalised map showing the distribution of late Cenozoic basins in California, in relation to the San Andreas transform fault system. (Adapted from Blake *et al.*, 1978 and others.)

distances from near-shore, or even continental, deposits to deep-sea turbidites and, because of localised synsedimentary tectonic movements, thicknesses will be subject to rapid variation. Geothermal gradients, similarly, will vary widely.

The supreme examples of such basins, especially in relation to their petroleum interest, must be those of central and southern California (Fig. 3.11) (e.g. Crowell, 1974; Blake *et al.*, 1978; Nardin & Henyey, 1978). Indeed, the Los Angeles Basin is the richest in the world in terms of petroleum yield per unit volume of sediment. As many as 30 small, more or less separated onshore and offshore basins of Tertiary to Recent age can be identified, all of them in some way related to the San Andreas Fault system. Their sediments were largely derived from intervening uplifts, and include almost the entire range of clastic deposits together

Chapter 3

with some carbonates. California today is situated adjacent to an area of upwelling of mineral-rich oceanic waters, leading to high organic productivity, and this factor may have contributed to the high organic content of the Cenozoic sediments. Petroleum is obtained from reservoirs ranging from coarse clastics to turbidites, and in anticlinal, fault-closed and stratigraphic traps in a great variety of local tectonic environments. There can be no doubt that the whole process from sedimentation, through trap formation to petroleum generation, migration and accumulation, is continuing up to the present time.

3.5 Intraplate basins

Sedimentary basins occur within crustal plates, entirely within the continental shields (intracratonic), superimposed upon or within fold-belts, at the margins between continents and oceans and within the oceanic realm (Section 4.4). It has already been pointed out that, strictly speaking, intraplate basins evolve through causes not directly attributable to plate tectonics: however, a consideration of them shows that some of the environments were created initially through such processes, and that plate tectonic influences may still be apparent. It is therefore appropriate that we should review them briefly.

INTRACRATONIC BASINS

The basins developed within shield areas are often surprisingly circular in outline. They may be long-lived and sedimentation can keep pace with subsidence (e.g. Sloss & Speed, 1974), so that not only do shallow-water deposits predominate but interval isopachs tend to reflect the shape of the basin. The nature of the sediments, and hence petroleum potential, will depend upon marine connections but, where they are well established, organic matter may be concentrated. Geothermal gradients tend to be approximately normal for shield areas, and traps are either related to basement swells or ridges, or are stratigraphic. Some intracratonic basins, such as those in North Africa, extend to the continental margin and, in such an environment, may later be partially involved in continental collision and orogenesis.

Outstanding examples are the Moscow Basin, where the sediments are largely continental and petroleum interest is low, and the Williston Basin in the north central United States and Canada (Fig. 3.12). The latter persisted from the early Palaeozoic to the Tertiary and the sedimentary thickness reaches 4 km. Marine connections were good and the petroleum generated within the basin has accumulated primarily over gentle basement ridges.

Petroleum: the Sedimentary Basin

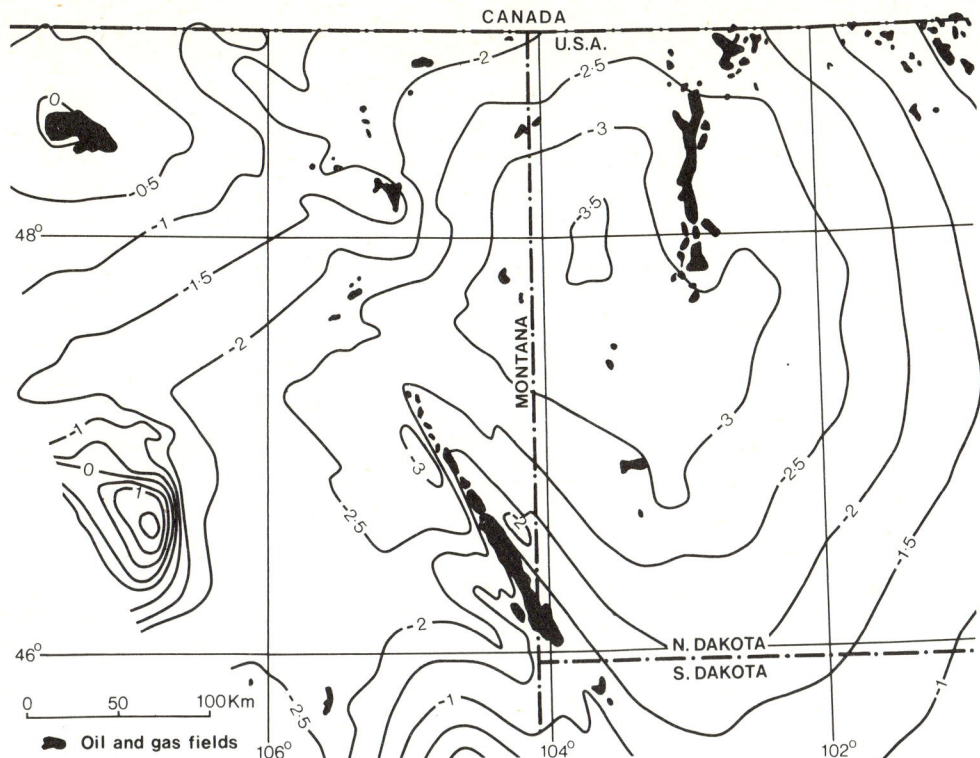

Fig. 3.12. The Williston Basin. Generalised structure contours (in thousands of feet from sea-level) on the Cretaceous, indicating the broad relationship of oil and gas fields to positive relative upwarps. (Simplified after Harding & Lowell, 1979.)

FOLD-BELT BASINS

Included here are both those basins which have developed from a formerly stable 'block' within the fold-belt and have later subsided, and those which have been superimposed on it, possibly over a tectonic sag. Examples of the former are the central European Pannonian Basin and the present day Black Sea (Degens & Ross, 1974) whilst the latter include the Vienna and Maracaibo Basins. Both categories are treated together because, as far as petroleum is concerned, they appear to have somewhat similar characteristics. They are rather restricted in area, and appear to have subsided and evolved rapidly. If an abundant supply of sediment is available, then the sediments may be coarsely clastic, even molassic, and shallow-water environments will predominate (Vienna Basin). If, on the other hand, they are relatively starved, then considerable water depths will remain (Black Sea). Marine connections may be rather restricted, leading to organic enrichment, in which case the basin may be very petroliferous. Structures may be compressional, being inherited from preceding orogeny, or purely tensional: the latter are perhaps the more common.

Chapter 3

INACTIVE CONTINENTAL MARGINS

(These continue the 'rift-drift' sequence that was considered earlier.) Once equilibrium has been reached following the splitting of a continental mass, then stable continental margins will persist until they are involved in a continental collision. This may be very much later, so that the continent-ocean margin can remain in this stable state for a long period of time. Basin development is through the prograding of sediments beyond the continental edge and, generally but not always, broad subsidence of a shallow marine shelf within the margin. The continental shelf may also be subject to important gentle differential vertical movements within the basement, and, as for example in Gabon and Angola (Brink, 1974; Brognon & Verrier, 1966) (Fig. 3.13), the shelf sediments commonly

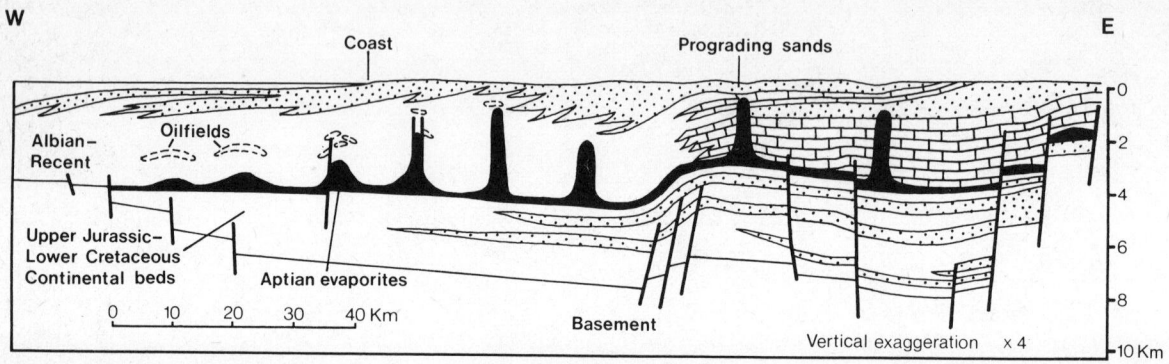

Fig. 3.13. Cross-section of the Gabon Basin, a typical inactive margin inherited from a former rift basin. Note the basal continental clastics with early evaporites and shallow marine deposits of the rift phase, overlain by oceanwards prograding beds of the inactive margin phase. (After Brink, 1974.)

overlie block-faulted, continental and early marine deposits inherited from the rifting stages of development. If evaporites are present, then diapiric structures may be expected. Inactive, or formerly inactive, margins are extremely important from the point of view of petroleum and a high proportion of the world's reserves are found in them: one has only to note that the Middle East Basin, extending from Saudi Arabia into Iran, evolved at an inactive margin until it was subjected to continental collision in the Late Tertiary (Figs 3.14 and 3.15).

The wide continental shelves reflect an aspect of plate tectonics not yet mentioned. They are, of course, very susceptible to marine transgressions and regressions, one cause being variations in the rate of sea-floor spreading: the argument is as follows, that rapid spreading gives rise to more elevated mid-ocean ridges, the displaced water resulting in a higher sea-level and transgression over the continental margins (Fig. 3.16). During slower spreading, the converse applies (e.g. Rona, 1973; Pitman, 1978). Although such sea-level variations are worldwide, their effects are particularly noticeable on the stable continental shelves, and they are extremely important from the point of view of petroleum. Transgression

Petroleum: the Sedimentary Basin

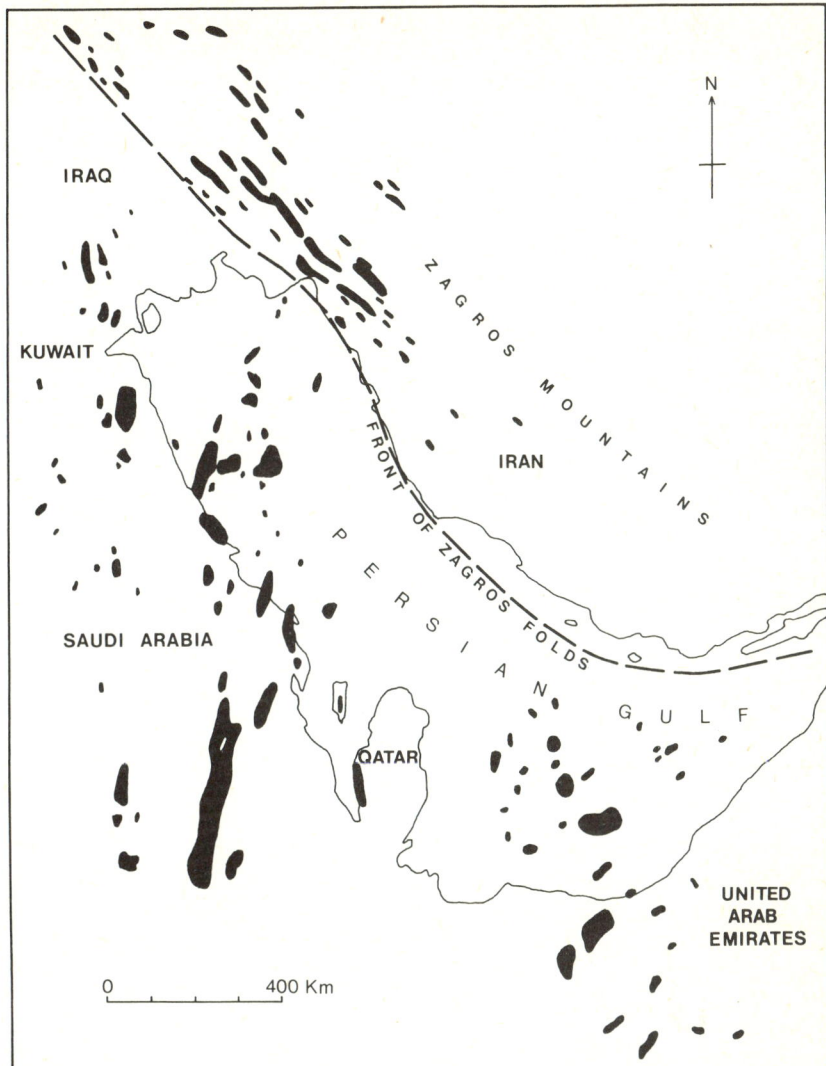

Fig. 3.14. Oilfields of the Middle East Basin. Elongations of the fields suggest their origin: those trending approximately north-south in Arabia are predominantly in structures related to basement block movements; rounded outlines of fields in the Persian Gulf are related to salt structures; north-west—south-east elongated fields in Iran represent Mesozoic oil distributed into anticlines of the Neogene Zagros foldbelt.

results in deeper water sedimentation and the accumulation of potential source rocks: shelf seas are often rich in organic matter, particularly if the continental margin is associated with oceanic upwelling. Regression leads to the accumulation of widespread sands or carbonates, perhaps reefal, and may culminate in lagoonal or sabkha evaporite precipitation. The interleaving of source rock, reservoir and cap-rock can thus be established: these favourable circumstances may be enhanced by a lateral juxtaposition of deeper water intrashelf depressions and shallow water 'highs', which can also form very broad traps. It was indeed this combination of

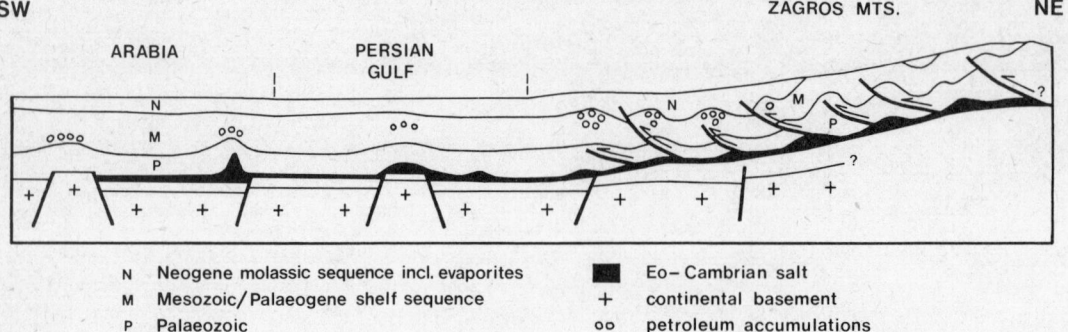

N	Neogene molassic sequence incl. evaporites	■	Eo–Cambrian salt
M	Mesozoic/Palaeogene shelf sequence	+	continental basement
P	Palaeozoic	oo	petroleum accumulations

Fig. 3.15. Diagrammatic sketch section of the regional structure and oilfields in the Middle East Basin.

factors that is believed to have been the reason for the abundance of petroleum in the Middle East.

Sediment progradation over continental margins has, in recent years, become more widely understood, in no small measure due to the applications of seismic techniques to stratigraphy (e.g. Payton, 1977). The continental slopes have, as yet, been little explored with the drill, and they may perhaps hold out the greatest hope for significant future discoveries, largely in stratigraphically controlled traps. However, one particular such environment has already proved to be very important with respect to petroleum.

Deltas prograding over the continental margin off the mouths of large rivers contain large reserves (Fig. 3.17). They result, in effect, in large-scale regressive sedimentary sequences, which display a fascinating association of interrelated phenomena (e.g. Chapman, 1973). Continuous deep-water shales at the base of the prograding sequence may retain much

Fig. 3.16. Diagrams showing the possible worldwide effect of crustal spreading rate on sea-level.

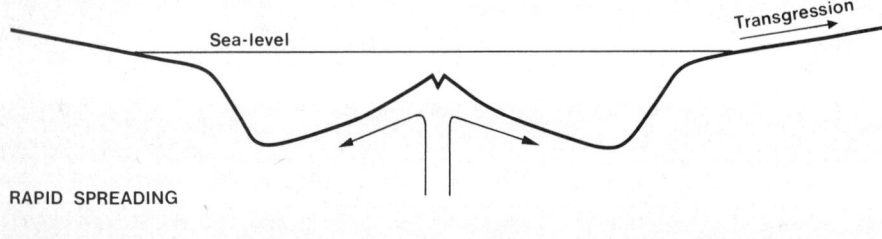

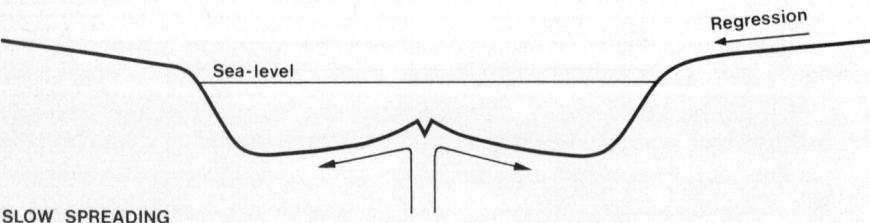

Petroleum: the Sedimentary Basin

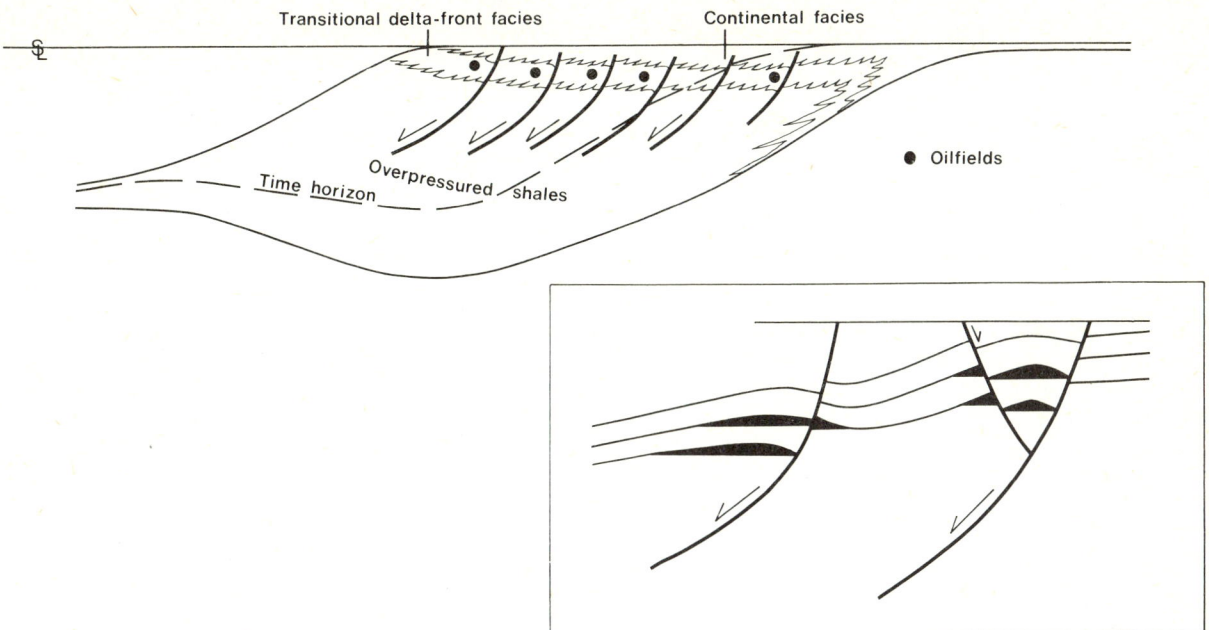

Fig. 3.17. The habitat of oilfields in major deltas. Inset: the cross-sectional relationship of growth-faults, roll-over anticlines and petroleum traps.

of their original interstitial water and are commonly under-compacted and overpressured (as compared with a hydrostatic head). As such they not only act as geothermal heat sinks but also cause instability in the overburden: this results in down-to-the-ocean slumping, expressed as synsedimentary normal faulting, and the development of associated 'roll-over' anticlines. We thus have the conditions for the generation of petroleum and its accumulation, generally in the prograded delta-front sands sealed by minor transgressive shales, in adequate but often complex structural traps. The Mississippi and Niger Deltas are the best known examples (e.g. Harding & Lowell, 1979; Weber & Daukoru, 1975; Evamy *et al.*, 1978).

A rather special situation is created when a formerly inactive shelf is involved in an interplate continental collision. A fore-deep, or molasse, basin may be developed in front of the resulting orogenic fold-belt, and may itself, at least in part, be involved in the later stages of this folding. Well-known examples, on either side of the European Alps, are the Neogene Po Valley to the south and the Molasse Basin to the north (e.g. Van Houten, 1974): the Siwalik trough on the southern margin of the Himalaya is another. Potential reservoirs and traps are commonly well developed, but the presence or absence of petroleum will depend primarily on whether or not the sea invaded the basin and, if so, whether or not a sufficient thickness of source rocks (and cap-rocks) could accumulate. It is important not to confuse petroleum endemic to the underlying, pre-orogenic shelf sequence, as is well exemplified in south-west Iran (Fig.

Chapter 3

3.13), with any that may be related *sensu stricto* to the synorogenic basin, which can be barren. No indigenous petroleum is present in Iran in the thick Upper Miocene-Pliocene molassic deposits related to the foredeep of the Zagros orogene: the oil and gas in the Zagros anticlines are derived from the Mesozoic shelf beds, and the former situation is preserved in Arabia, which has not been affected by the compressional folding.

3.6 Conclusions

This rapid and sketchy review of the plate tectonic development of sedimentary basins, and of the major habitats of petroleum, should serve to emphasise that aspects of the Earth's crust as diverse as plate tectonic processes and petroleum, in fact may be merely different manifestations of a whole spectrum of processes in the outer layers of our planet. With this in mind, the next chapter will examine in more detail the controls on petroleum accumulations, and see what impact plate tectonics may have on our understanding of them.

References

Baker, B.H., Mohr, P.A. & Williams, L.A.J. (1972) *Geology of the Eastern Rift System of Africa.* Geol. Soc. Am. Spec. Pap. 136, 67 pp.

Bally, A.W. (1975) A geodynamic Scenario for hydrocarbon occurrences. *Proc. 9th World Petrol. Congr. Tokyo,* Vol. 2, pp. 33-34.

Beaumont, C. (1978) The evolution of sedimentary basins on a viscoelastic lithosphere: theory and examples. *Geophys. J. R. astr. Soc.,* **55,** 471-497.

Blake, M.C., Campbell, R.H., Dibblee, T.W., Howell, D.G., Nilsen, T.H., Normark, W.R., Vedder, J.C. & Silver, E.A. (1978) Neogene basin formation in relation to plate-tectonic evolution of San Andreas Fault system, California. *Bull. Am. Assoc. Petrol. Geol.,* **62**, 344-372.

Brennand, T.P. & van Veen, F.R. (1975) The Auk Oil-Field. In *Petroleum and the Continental Shelf of North West Europe, Vol. 1 Geology* (Ed. A.W. Woodland) Applied Science Publishers, London, pp. 275-284.

Brink, A.H. (1974) Petroleum Geology of Gabon Basin. *Bull. Am. Assoc. Petrol. Geol.,* **58**, 216-235.

Brognon, G.P. & Verrier, G.R. (1966) Oil and geology in Cuanza Basin of Angola. *Bull. Am. Assoc. Petrol. Geol.,* **50**, 108-158.

Chapman, R.E. (1973) *Petroleum Geology: A Concise Study* Elsevier, Amsterdam, London and New York, 304 pp.

Crowell, J.C. (1974) Origin of late Cainozoic basins in southern California. In *Tectonics and Sedimentation* (Ed. W.R. Dickinson) Soc. Econ. Pal. & Min. Spec. Publ. 22, pp. 190-204.

Curray, J.R., Moore, D.G., Lawyer, L.A., Emmel, F.J., Raitt, R.W., Henry, M. & Kieckhefer, R. (1979) Tectonics of the Andaman Sea and Burma. In *Geological and Geophysical Investigations of Continental Margins.* (Eds J.S. Watkins, L. Montadert & P.W. Dickerson). Am. Assoc. Petrol. Geol., Mem. 29, pp. 189-198.

de Coster, G.L. (1975) The geology of the central and south Sumatran basins. *Proc. 3rd Ann. Convention (1974). Indonesian Petrol. Ass.* Jakarta, pp. 77-111.

Degens, E.T. & Ross, D.A. (Eds) (1974) *The Black Sea—Geology, Chemistry and Biology.* Am. Assoc. Petrol. Geol., Mem. 20, 633 pp.

Dickinson, W.R. & Seely, D.R. (1979) Structure and stratigraphy of forearc regions. *Bull. Am. Assoc. Petrol. Geol.*, **63**, 2-31.

Evamy, D.D., Haremboure, J., Kamerling, P., Knaap, W.A., Molloy, F.A. & Rowlands, P.H. (1978) Hydrocarbon habitat of Tertiary Niger Delta. *Bull. Am. Assoc. Petrol. Geol.*, **62**, 1-39.

Feininger, T. (1975) Origin of oil in the Oriente of Ecuador. *Bull. Am. Assoc. Petrol. Geol.*, **59**, 1166-75.

Fisher, M.A. & Magoon, L.B. (1978) Geologic Framework of Lower Cook Inlet, Alaska. *Bull. Am. Assoc. Petrol. Geol.*, **62**, 373-402.

Glennis, K.W. (1972) Permian Rotliegendes of north-west Europe interpreted in light of modern desert sedimentation studies. *Bull. Am. Assoc. Petrol. Geol.*, **56**, 1048-71.

Harding, T.P. & Lowell, J.D. (1979) Structural styles, their plate-tectonic habitats, and hydrocarbon traps in petroleum provinces. *Bull. Am. Assoc. Petrol. Geol.*, **63**, 1016-1058.

Hassan, F. & El-Dashlouty, S. (1970) Miocene evaporites of Gulf of Suez region and their significance. *Bull. Am. Assoc. Petrol. Geol.*, **54**, 1686-96.

Hay, J.T.C. (1978) Structural development in the northern North Sea. *Jl. Petrol. Geol.*, **1**, 65-77.

Kay, M. (1951) *North American Geosynclines.* Mem. Geol. Soc. Am., p. 48.

Kent, P.E. (1977) The Mesozoic developments of aseismic continental shelves. *J. geol. Soc. Lond.*, **134**, 1-18.

Kirschner, C.E. & Lyon, C.A. (1973) Stratigraphic and tectonic development of Cook Inlet petroleum province. In *Arctic Geology* (Ed. M.G. Pitcher) Am. Assoc. Petrol. Geol., Mem. 19, pp. 396-407.

Klemme, H.D. (1975) Giant oilfields related to their geologic setting—a possible guide to exploration. *Bull. Can. Petrol. Geol.*, **23**, 30-66.

Lonsdale, P. (1978) Ecuadorian subduction system. *Bull. Am. Assoc. Petrol. Geol.*, **62**, 2454-77.

Lowell, J.D., Genik, G.J., Nelson, T.H. & Tucker, P.M. (1975) Petroleum and Plate Tectonics of the southern Red Sea. In *Petroleum and Global Tectonics* (Eds A.G. Fischer & S. Judson) Princeton University Press, Princeton, pp. 129-153.

Nardin, T.R. & Henyey, T.L. (1978) Pliocene-Pleistocene diastrophism of Santa Monica and San Pedro Shelves, California Continental Borderland. *Bull. Am. Assoc. Petrol. Geol.*, **62**, 247-272.

Osmaston, M.F. (1977) Some fundamental aspects of plate tectonics bearing on hydrocarbon location. In *Developments in Petroleum Geology—1* (Ed. G.D. Hobson) Applied Science Publishers, London, pp. 1-52.

Payton, C.E. (Ed.) (1977) *Seismic Stratigraphy—Applications to Hydrocarbon Exploration.* Amer. Assoc. Petrol. Geol., Mem. 26, 516 pp.

Pegrum, R.M. & Mounteney, N. (1978) Rift basins flanking North Atlantic Ocean and their relation to North Sea area. *Bull. Am. Assoc. Petrol. Geol.*, **62**, 419-441.

Pennington, J.J. (1975) The geology of the Argyll Field. In *Petroleum and the Continental Shelf of North West Europe Vol. 1 Geology* (Ed. A.W. Woodland) Applied Science Publishers, London, pp. 285-294.

Pitman, W.C. (1978) Relationship between eustacy and stratigraphic sequences of passive margins. *Bull. geol. Soc. Am.*, **89**, 1389-1403.

Robson, D.A. (1971) The structure of the Gulf of Suez (Clysmic) rift, with special reference to the eastern side. *J. geol. Soc., Lond.*, **127**, 247-276.

Rona, P.A. (1973) Relations between rates of sediment accumulation on continental shelves, sea-floor spreading and eustacy inferred from the central North Atlantic. *Bull. Geol. Soc. Am.*, **84**, 2851-72.

Seely, D.R. (1979) The evolution of structural highs bordering major forearc basins. In *Geological and Geophysical Investigations of Continental Martins.* (Eds J.S. Watkins, L. Montadert & P.W. Dickinson) Am. Assoc. Petrol. Geol., Mem. 29, pp. 245-260.

Selley, R.C. (1976) The habitat of North Sea Oil. *Proc. Geol. Assoc.*, **87**, 359-88.

Chapter 3

Sloss, L.L. & Speed, R.C. (1974) Relationships of cratonic and continental margin tectonic episodes. In *Tectonics and Sedimentation.* (Ed. W.R. Dickinson) Soc. Econ. Pal. & Min. Spec. Publ. 22, pp. 98-119.

Stoneley, R. (1967) The structural development of the Gulf of Alaska sedimentary province in southern Alaska. *Jl. geol. Soc. Lond.,* **123**, 25-57.

Stoneley, R. (1969) Sedimentary thicknesses in orogenic belts. In *Time and Place in Orogeny.* (Eds P.E. Kent *et al.*) London (Geological Society), pp. 215-238.

Thompson, T.L. (1976) Plate tectonics in oil and gas exploration of continental margins. *Bull. Am. Assoc. Petrol. Geol.,* **60**, 1463-1501.

Travis, R.B., Gonzales, G. & Pardo, A. (1976) Hydrocarbon potential of coastal basins of Peru. In *Circum-Pacific Energy and Mineral Resources.* (Eds M.T. Halbouty, J.C. Maher & H.M. Lian.) Am. Assoc. Petrol. Geol., Mem. 25, pp. 177-188.

Van Houten, F.B. (1974) Northern Alpine Molasse and similar sequences of southern Europe. In *Modern and Ancient Geosynclinal Sedimentation* (Eds R.H. Dott & R.H. Shaver.) Soc. Econ. Pal. & Min. Spec. Publ. 19, pp. 260-273.

Walmsley, P.J. (1975) The Forties Field. In *Petroleum and the Continental Shelf of North West Europe, Vol. 1, Geology* (Ed. A.W. Woodland) Applied Science Publishers, London, pp. 477-486.

Weber, K.J. & Daukoru, E. (1975) Petroleum Geology of the Niger Delta. *Proc. 9th World Petrol. Congr. Tokyo* Vol. 2, pp. 209-221.

Woodland, A.W. (1975) *Petroleum and the Continental Shelf of Northwest Europe, Vol. 1, Geology* Applied Science Publishers, London, 501 pp.

Ziegler, P.A. (1975) Geologic evolution of North Sea and its tectonic framework. *Bull. Am. Assoc. Petrol. Geol.,* **59**, 1073-97.

Ziegler, P.A. (1977) Geology and hydrocarbon provinces of the North Sea. *GeoJournal,* **1**, 7-32.

Zuniga y Rivero, F., Pardo, A., Valdivia, H. & Velarde, P. (1976) Hydrocarbon Potential of Amazon Basins of Colombia, Ecuador and Peru. In *Circum-Pacific Energy and Mineral Resources* (Eds M.T. Halbouty, J.C. Maher & H.M. Lian) Am. Assoc. Petrol. Geol., Mem. 25, pp. 339-348.

Chapter 4 Petroleum: Entrapment and Conclusions

R.J. BAILEY & R. STONELEY

4.1 Reservoirs*

A reservoir rock has two essential properties: it must be porous to contain oil or gas, and it must be permeable so that fluids can migrate through it, and thence into a borehole. *Porosity* is merely the proportion of the void space to the bulk volume of the rock; it is commonly expressed as a percentage (ϕ). Within some rocks there will be void spaces that are isolated from each other: they are not interconnected, and are of no use to us. A further restriction is that, in most reservoirs, there is a thin film of water adhering to the walls of the pores, held there by strong interfacial tensions (Fig. 4.1). This water is immovable and it limits the amount of petroleum that the rock can contain. The percentage of the pore space that this water occupies is known as the *water saturation* (S_w). The remaining space that the oil, say, can occupy is referred to as the *effective porosity* to oil [$\phi(1-S_w)$].

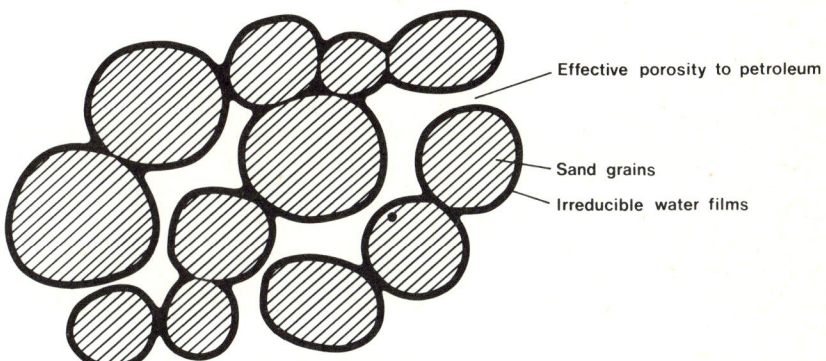

Fig. 4.1. The effect of irreducible water films on the effective porosity of a sandstone.

In a clastic rock porosity is independent of the size of the grains: thus, the porosity of a pile of billiard balls is the same as for a similarly stacked pile of cannon balls. However, the smaller the grains the tighter the pore spaces and, in a fine-grained rock, the higher will be the proportion of the porosity occupied by the irreducible water films. Hence the effective porosity to oil will be to a greater or lesser extent dependent upon pore

*Sections marked thus are written by R. Stoneley.

size: although exceptions are known, a very fine-grained chalk, for example, may be quite porous to water as a continuous phase, but have virtually zero effective porosity to oil. Our ideal reservoir rock, therefore, will be coarse-grained, and will have maximum porosity if the grains are of equal-size and spherical.

Permeability is a measure, expressed in darcies, of the ability of the rock to allow a fluid to pass through it. It is measured experimentally in the laboratory and, like porosity, is a property of the rock: if only one fluid is present, it is independent of the nature of that fluid. Just as the immovable water films reduce the effective porosity to oil, so they also affect the permeability and an *effective permeability* to oil or gas can likewise be measured. In the same way that grain size affects the effective porosity, the finer-grained reservoirs also have a lower effective permeability to petroleum. A strongly adverse effect on both parameters is provided by poor grain sorting: the smaller particles not only fill much of the pore space between the larger grains, but also greatly increase the surface area to which water adheres, hence drastically reducing both effective porosity and permeability.

These conditions refer most to sandstones, but the same physical principles apply to the porosity developed in carbonates. These two broad lithological types are by far the most important as petroleum reservoirs: although production is obtained locally from fractured shales and jointed or weathered igneous rocks, they would not normally represent an exploration objective and may be ignored for the present purposes.

SANDSTONE RESERVOIRS

For reservoir purposes, the optimum sandstone will consist of coarse, well-sorted, rounded grains, cemented only sufficiently enough to give the rock an essential cohesion; the thicker and more isotropic the reservoir is the better and, in order to allow migration over a wide gathering area and to provide good pressure communication, it should be laterally extensive. These ideal qualities are not, in practice, found very often. Consider the environments in which clean sands accumulate (Fig. 4.2): in dunes, principally in deserts; in certain fluvial, lacustrine and deltaic settings; in beaches and offshore bars; and in deep-sea grain-flow deposits. Most of these environments are, at any moment in time, restricted areally, and we are dependent on their gradual migration laterally to provide lithological continuity—through such processes as channel meandering and marine transgression and regression. As a result, many sandstone reservoirs are limited in vertical and areal extent, or are affected by inhomogeneities which can greatly influence production performance. Let us look at three contrasting sequences in which sandstones are important reservoirs.

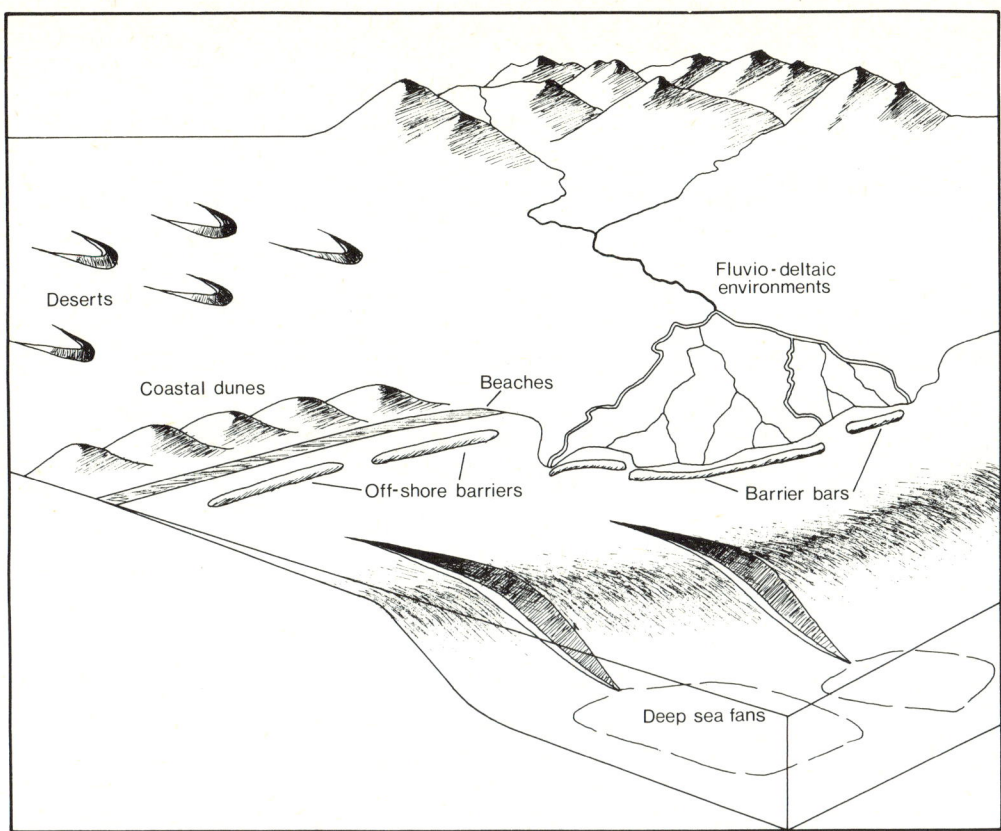

Fig. 4.2. The principal sedimentary environments of reservoir sands.

The Niger Delta comprises a vast clastic sequence which is prograding oceanwards, so that deep-sea clays are overlain by a diachronous delta front and deltaic deposits, which in turn pass upwards into mainly sandy continental beds (see Fig. 3.17) (e.g. Weber and Daukoru, 1975; Evamy *et al.*, 1978). Many minor transgressions and regressions have caused the transition zone to be divided into separate sandstone bodies, several metres thick, each of which passes laterally southwards into marine clays and landwards into the continental sands. Thus, in a single oilfield there can be 50 or more individual sandstone reservoirs. The situation is even more complicated than this, as drilling within individual fields has shown that each sand can be broken into many, laterally discontinuous bodies, representing very localised depositional environments (Fig. 4.3) (e.g. Weber, 1971): the problems of detailed correlation and of controlling production are severe indeed, and they do not become apparent until detailed data are available!

Sandstones of stable shelf environments are exemplified by the Aptian to Cenomanian sands, which spread north-eastwards from the Arabian

Chapter 4

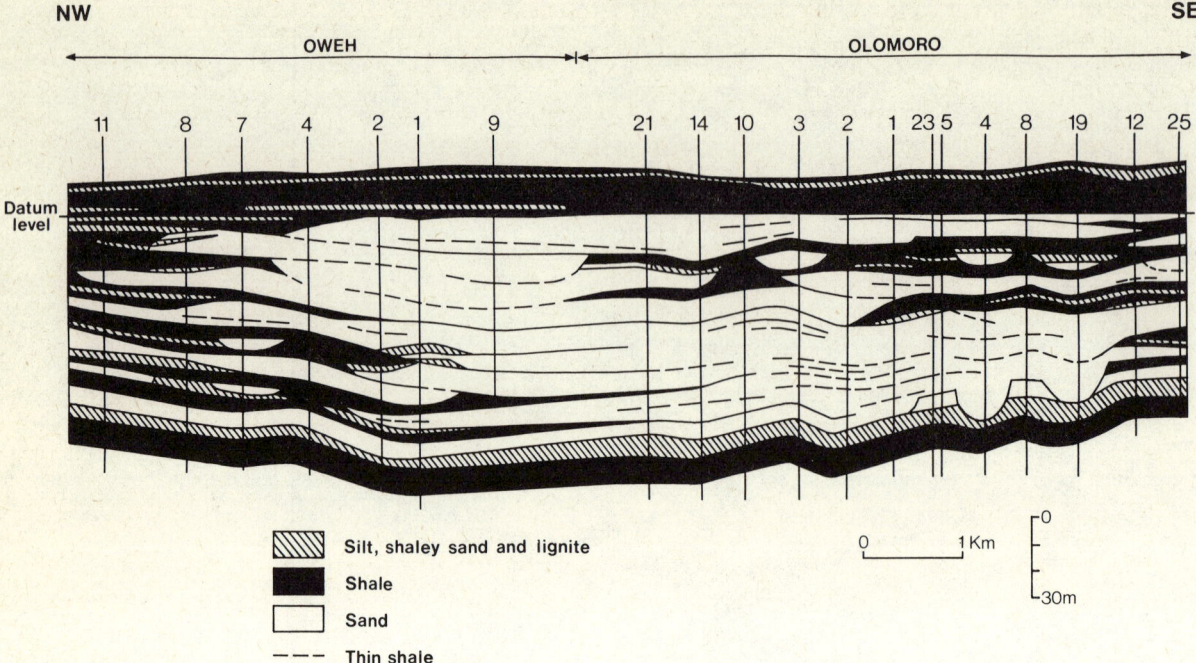

Fig. 4.3. Variations in the distribution of reservoir sands in a part of the paralic sequence in the Oweh and Olomoro Fields in the Niger Delta (after Weber, 1971).

Shield to the oilfield areas of Iraq, Kuwait and Saudi Arabia (e.g. Adasani, 1965; Schlumberger, 1975). Three spreads of sand, each reaching up to 300 m or more in thickness, were deposited, for example in the Kuwait area, in near-shore environments; although some argillaceous and poorer sand intervals are present, porosities and permeabilities in the better sorted sands are exceptionally high, exceeding 30% and 4 darcies. Shoreline oscillations appear to have ensured extensive lateral continuity: this is one more of the factors that make the Middle East Basin so prolific!

For the third example, the Permian Rotliegendes Formation of the southern North Sea region may be cited (e.g. Glennie, 1972). Aeolian sands predominate over much of the area and, due to the shifting of desert dunes, can form almost uniform accumulations more than 100 m thick; porosities normally range up to about 20%, and permeabilities up to 2 darcies have been measured.

CARBONATE RESERVOIRS

Carbonate rocks are characteristic of shallow-water seas, and become important principally on more or less stable continental shelves. Primary porosity is intergranular in oolitic or clastic limestones, and is also found between and within the skeletons of such organisms as the reef-building corals. However, carbonates are very susceptible to cementation and to solution, mineral alteration and reprecipitation through percolating waters. Thus secondary porosity becomes important—selective leaching, the 13% volume reduction on the alteration of calcite to dolomite, the leaching of secondarily deposited anhydrite, etc. Although such porosity points to the slow percolation of water over geological time, much of it may be ineffective in terms of petroleum. Carbonate porosities and permeabilities are thus normally lower than in sandstones, although, since a limestone commonly forms a firmly cemented and rigid body, it may be more liable to fracturing thus creating channels of high productivity.

One of the most outstanding examples of a thick petroliferous carbonate sequence is again provided by the Middle East Basin, extending from Syria and Turkey to Oman and from Iran into the interior of Saudi Arabia. This vast area was covered by an essentially shallow shelf sea from the Permian to the Miocene and carbonates predominante in a wide range of facies. Interbeds include argillaceous rocks which have provided the source for the abundant petroleum, evaporites which in some cases constitute cap-rocks, and the sandstones derived from the Arabian shield (pp. 75-6). The facies variations stem from fluctuations of sea-level and from differential basement subsidence, which is also responsible for some of the structural traps. Porosity is principally secondary, but nevertheless it tends to be developed especially in the originally coarser deposits. Some primary porosity remains and rudist reefs are important locally. In Iran and Iraq, where the sequence has been involved in the Neogene Zagros folding, extensive fracturing can give very high production rates.

Since reefs are especially important in localising porosity, it is appropriate also to refer to the widespread Palaeozoic shelf sequence of western Canada. Reefs are particularly well developed in the Devonian and contain a high proportion of the region's oil and gas reserves. Most of the porosity is secondary, in coarsely dolomitised biohermal and biostromal 'build-ups', whose distribution is related to localised depressions within the shelf. (Brief summary and references in Selley, 1978, p. 212 *et seq.*)

RESERVOIRS AND PLATE TECTONICS

What has plate tectonics to do with all this? The immediate answer, when looking at an individual reservoir, may be 'very little'! Such considerations will, however, influence our thinking both when considering the type and general form of reservoirs on a regional scale, and also perhaps in predicting their distribution within a particular basin.

Clean sands require an abundant source of detritus and a high-energy depositional environment. The development of thick carbonates, on the other hand, needs rather stable tectonic conditions, persisting over a wide area and for a considerable period of time: initial formation is in generally shallow-water environments, although reworking by submarine currents can lead to redeposition in deeper water. Let us consider the plate tectonic situations for these conditions (ref. also Dickinson & Suczek, 1979).

In the early stages of the 'rift-drift' sequence, uplift and block-faulting will ensure a source of clastic debris; this may be deposited rapidly in the rift under continental conditions, some of it probably in high-energy environments. After marine incursion and during early continental separation, the basin may still receive detritus from the margins, but it is likely to be concentrated near the flanks of the basin, either near shore or possibly in small deltas; its distribution may be controlled by continued block-faulting. In areas sheltered from clastic influx, and in general provided that the sea is warm, carbonates and even evaporites may begin to accumulate. Later in the sequence, a continental shelf at an inactive margin will be the site of deposition of both carbonates and clastic rocks, including clays, depending on the climatic conditions, on the supply of sediment from the land, and on the development of localised irregularities in the floor of the shelf; the detrital sediments may prograde beyond the shelf edge and, under the influence of submarine currents, give rise to well-sorted reservoirs in deeper waters. Thus, the 'rift-drift' sequence of events is liable to give rise to attractive reservoirs, both sandstone and carbonate, at most stages of its evolution. Trends, both structural and stratigraphical, are liable to be parallel to the original rift, except where influenced by transverse currents and by intracontinental tectonics, including perhaps the development of discordant aborted rifts (aulacogens).

Active continental margins, on the other hand, are, at least in the fore-arc regions, notoriously poor in reservoir rocks. Clastic deposits tend to be immature, poorly sorted and irregularly distributed: turbidites are common, and there is little opportunity for carbonates to form. In the inter-arc and back-arc basins things may improve: clastic sediments tend to be of shallower water environments and can become better sorted; limestones may form in the more sheltered areas. Again, trends tend to be longitudinal.

In transform fault-related and intracontinental basins, reservoirs are commonly well developed. However, they can be very varied, and their facies will be related to the local morphology of the basin and to the overall palaeogeographic setting. Generalisation is too dangerous to be of value in these instances.

Alternatively we can look at the relationship from the opposite point of view. In what palaeogeographical environments can we expect to find the reservoirs that we have discussed? Aeolian sands are typically the deposits of desert regions, which today occur predominantly from approximately 10 to 35° North and South, although in some interior continental masses they extend into higher latitudes. Variations evidently existed in the past, depending on global climates, glaciations and so forth, but we would expect to find palaeodeserts in relatively warm regions, generally within continents where there was an exposed source of coarse clastic detritus. Preservation may be in intracratonic or early rift basins, and occasionally interleaved with continental shelf deposits, where wind-blown sands may spread during episodes of marine regression.

Dunes are, of course, also developed along certain, especially regressive, coastlines. Preservation potential, however, is poor, since redistribution by fluvial action or by intermittent marine transgression is common. River action may redistribute desert sands, notably during flash floods, but fluvial sands can occur in almost all climatic regions. Thus any sequence of continental deposits can be expected to contain intervals of well-sorted and winnowed sands. We have only to consider the northern hemisphere Carboniferous Coal Measures, or indeed any of the examples of continental successions already cited, to see that this is true; however, being susceptible to sub-aerial exposure and to percolation by waters of varying pH, they may be subjected to post-depositional changes which can affect their reservoir properties. Again, such sequences can be particularly characteristic of the early deposits of intracratonic and rift basins, but they can occur in most epicontinental situations during periods of lowered sea-level. Continental sequences are also characteristic of synorogenic molassic basins, but are of less interest to petroleum since source and caprocks are commonly lacking.

Coastal deposits, beach, bar and barrier sands, are subject to many of the same considerations, and of course may be formed to a greater or lesser extent in almost any littoral environment. However, they can be expected to be most extensively developed over wide and rather stable shelves, where they may be reworked several times by wave and current action. We are, therefore, led primarily towards the inactive continental margins developed during the later stages of the 'rift-drift' sequence, to back-arc basins in a continental region, and perhaps to the flanks of intracratonic basins.

A special case of combined sandstone environments is offered by delta systems, in which possible reservoirs are developed in a number of local depositional settings. Deltas obviously require a large river draining an elevated hinterland, but the accumulation of their sediments can only take place to a significant thickness where these are deposited in a pre-existing or contemporaneously formed crustal depression. Such a situation is found at the edge of a continental shelf, where sedimentation can spread into the deeper waters beyond, and this is the basinal environment of the Mississippi and Niger Deltas. The steep margins of an intracontinental or rifted basin can provide similar conditions, especially where such margins are controlled by faulting: the Middle Jurassic deltaic deposits on the western flank of the North Sea Basin are an increasingly well-known example. In plate tectonic terms, although deltaic sequences could be expected to be found associated with most types of basin, the greatest thicknesses would be anticipated at inactive continental margins, whereas the fore-arc regions of destructive margins are unstable and less likely to provide a shallow environment for a sufficient length of time. Submarine grain flows can result in the almost instantaneous deposition of a sand body, well sorted and thick enough to form a locally adequate reservoir. In this case, a steep basin margin and a topographic depression of the sea floor would be required for its accumulation, such as could be present in most basinal environments beyond the shallow shelf edge.

Finally, as a gross generalisation, carbonates require almost exactly the opposite conditions to the grain-flow sand! Oceanic reefs, associated with sea-mounts, may be prominent amongst modern carbonates, but in view of their common isolation from clastic sedimentation they generally lack effective burial, as well as source and cap-rocks: they consequently have a low petroleum potential. Carbonate sequences of interest, on the other hand, are developed on relatively stable continental shelves, either as inactive margins or in more enclosed seas, such as are provided by intra-continental and sometimes back-arc and second-stage rift basins involving continental crust. Structures associated with the development of the basin, and which can often be predicted from plate tectonic considerations, may exert a control on the distribution of the various carbonate facies.

4.2 Petroleum traps*

The oil and gas generated through thermal alteration of the organic matter in a source rock are expelled into situations of lower fluid potential on compaction of the source rock. These situations are initially in permeable reservoir rocks where there is lateral pressure communication. The petroleum will now migrate through the reservoir in the direction of decreasing potential and, if possible, will escape to the Earth's surface

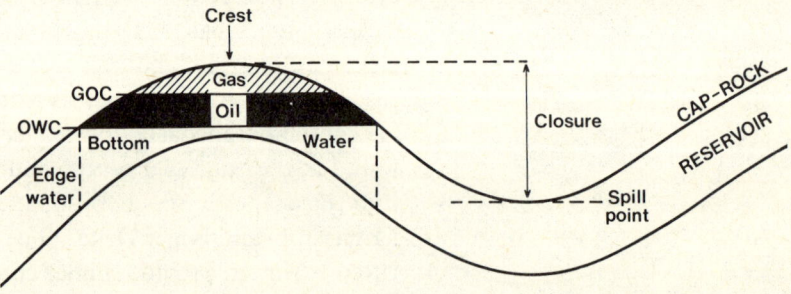

Fig. 4.4. Some basic parameters of a petroleum trap.

Petroleum: Entrapment and Conclusions

where it is lost. The concentrations of petroleum at this stage are very small and cannot be exploited. Sometimes, however, the migrating oil will reach a situation where its further migration is barred and there it will accumulate: from a physical point of view these trapping situations can be regarded as places of local minimum values in the fluid potential gradient along the path of migration.

Gas, being the lightest fluid present, will rise to the top of the trap, where it will overlie a layer of oil, if present (Fig. 4.4): of course, this can only occur within the effective porosity of the reservoir and provided that the reservoir is sealed by a cap-rock. Equally clearly, the location of such a trap in the subsurface must be one of the principal objectives of petroleum exploration.

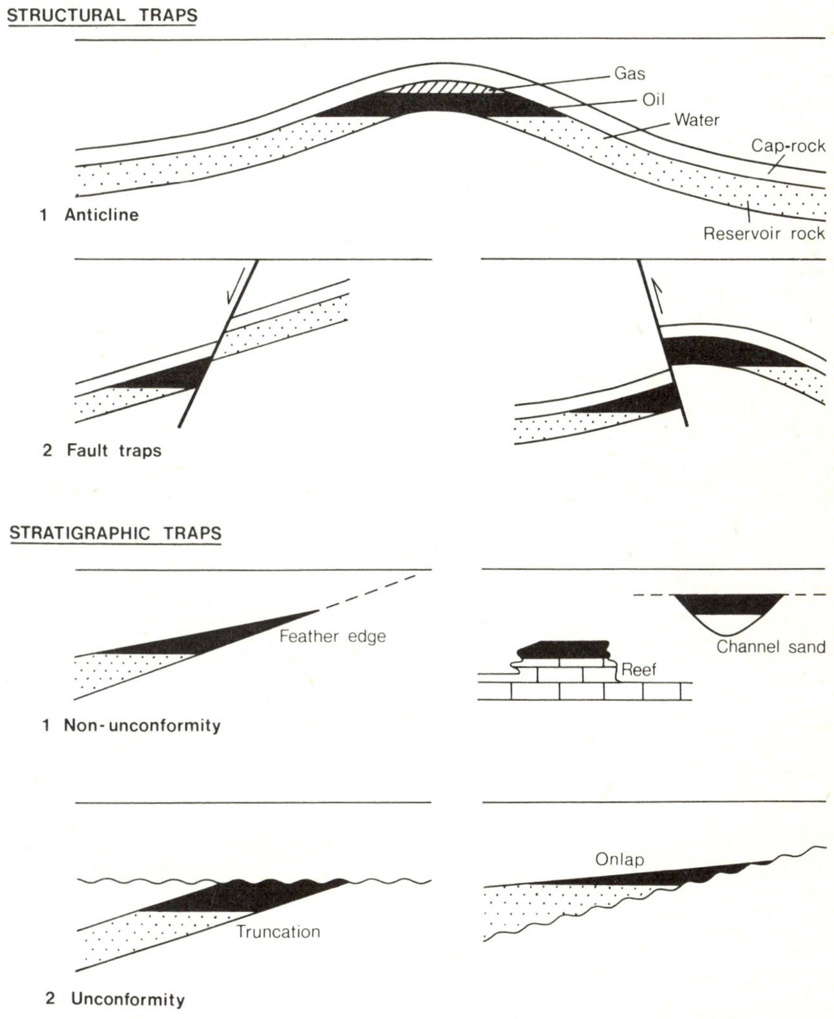

Fig. 4.5. Basic forms of petroleum traps.

Chapter 4

TYPES OF TRAP

Traps can be developed structurally through the action of tectonic processes or as a result of stratigraphic variations where structural control may amount to no more than a regional dip. They are classified accordingly (Fig. 4.5), although some traps are formed through a combination of the two overriding controls.

Structural traps

Structural traps, and in particular those developed in doubly plunging anticlines, are by far the most important in terms of both the total number of fields discovered and their proportion of total proven reserves (e.g. Moody, 1975); however, this may, to some extent, reflect the fact that they are the most easy to locate. Faulting can also produce traps, where a reservoir is juxtaposed against an impermeable stratum or where material introduced into the fault itself may act as a seal. Traps can be formed by both normal and reversed faults, provided of course that there is also closure in the third dimension which is usually perpendicular to our line of cross-section: the eight theoretical geometries and their trapping conditions are shown in Fig. 4.6.

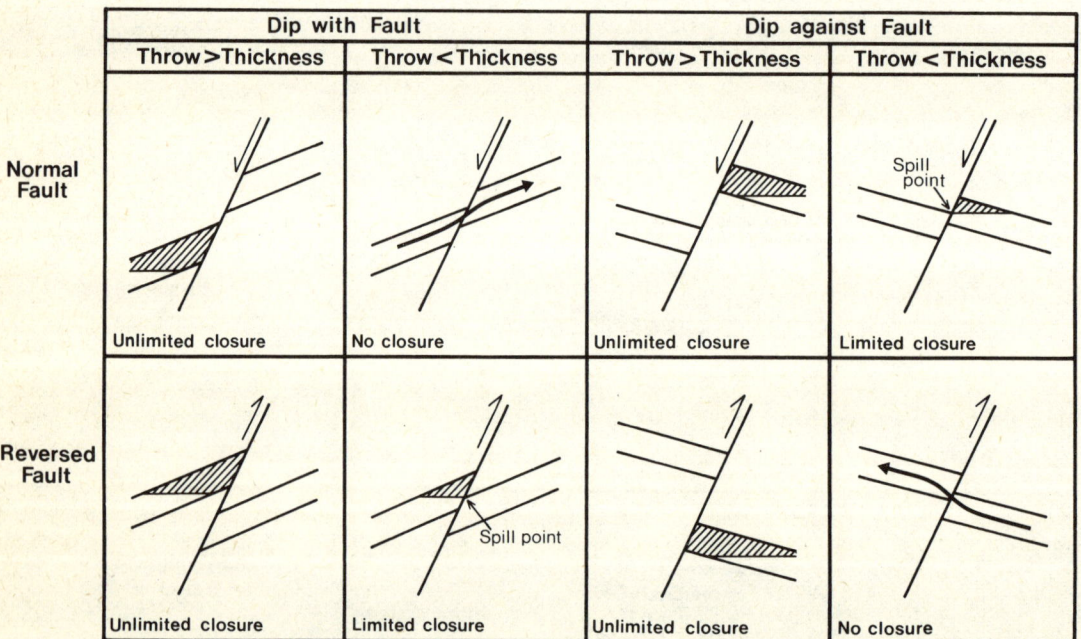

Fig. 4.6. The eight theoretical configurations of petroleum traps associated with faulting. Note that if the fault is sealing, all configurations would have unlimited closure.

Petroleum: Entrapment and Conclusions

Stratigraphic traps

Stratigraphic traps are conveniently classified according to whether or not they are associated with unconformities (e.g. Rittenhouse, 1972). In the former case, they may be developed either *above* or *below* the unconformity: above it we may find a reservoir sand or carbonate terminated laterally as a result of overlap or onlap of the overlying beds, or occasionally, if the unconformity surface is uneven, in localised valley fills or buried channels. Below an unconformity a trap may be formed by the truncation of an originally more steeply dipping reservoir bed, provided that the overlying sequence is impermeable. More esoteric circumstances can arise where porosity is developed in a buried topography, a classical example being the Augila Field in Libya where production is obtained from weathered and fractured basement granite (Williams, 1972). Stratigraphic traps, not associated with unconformity, are formed where a reservoir formation dies out laterally; and such situations can be developed locally in association with reefs, channel deposits, offshore bars, grain-flow bodies, etc.

Traps where there is a combination of trapping circumstances are exemplified by the Prudhoe Bay field of northern Alaska, in which a faulted anticline in basal Cretaceous and older beds is truncated by a younger Cretaceous sequence: the higher, impermeable rocks retain the oil within reservoirs of the older succession (Fig. 4.7). Finally, and for completeness, we may mention the *hydrodynamic* situation, where a flow

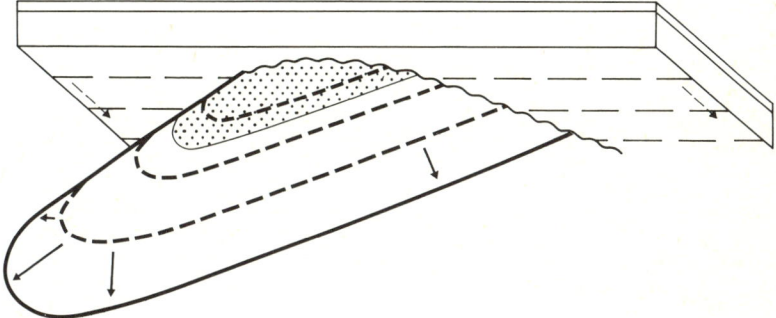

Fig. 4.7. A combination trap, where an anticline in an older series is truncated by a younger, uniformly dipping series. The trap at Prudhoe Bay in northern Alaska is of this type.

of water through a reservoir results in tilted levels of equal fluid potential: this will produce a tilted contact between a petroleum accumulation and the moving water, lateral displacement of that accumulation and, in an extreme case, could lead to accumulation where there is no complete structural closure.

TRAPS AND PLATE TECTONICS

Plate tectonics is likely to have its most obvious effect on the distribution of petroleum through the creation and preservation of structural

Chapter 4

traps, and we must look at the subject in greater detail. Of course, there is also a plate tectonic control on sedimentation and on the development of unconformities, as briefly considered in relation to the reservoirs themselves; this section, therefore, will be concerned primarily with structural traps, whilst not ignoring the importance of stratigraphical associations or the relationships between stratigraphy and structural development.

The problem can be looked at from the two points of view: in what tectonic situations are anticlinal and fault traps formed, and what kinds of trap would be expected in the various plate tectonic environments?

COMPRESSIONAL ANTICLINES

It is the superbly developed and exposed anticlines of the Zagros Mountains in south-west Iran, combined with the fact that where still buried they contain some of the world's most prolific oilfields, that must bring them to the fore in considerations of compressive folding (e.g. Lees, 1952; Falcon, 1969; Colmann-Sadd, 1978). These folds are largely concentric (parallel) and, below the centre of curvature where a space problem develops, there must be associated thrust-faulting and detachment from a 'basement' at depth (Fig. 4.8).

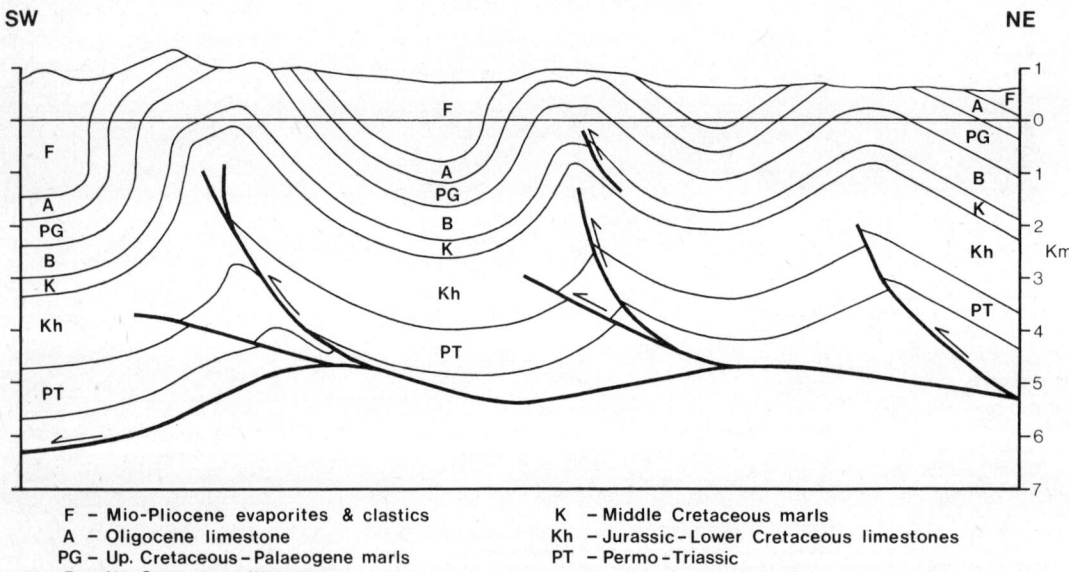

F – Mio-Pliocene evaporites & clastics
A – Oligocene limestone
PG – Up. Cretaceous – Palaeogene marls
B – Up. Cretaceous limestones
K – Middle Cretaceous marls
Kh – Jurassic – Lower Cretaceous limestones
PT – Permo-Triassic

Fig. 4.8. Concentric folds in the Zagros Mountains interpreted to depth. Maintenance of bed-length below the centre of curvature demands the presence of detachment thrusting at depth.

The nature of compressional folding will be governed principally by the lithological make-up of the folded sequence. Thus, whilst thick competent carbonate units in the Zagros give rise to essentially concentric folding, a less competent sequence can lead to less regular and often similar folds. The Middle Ground Shoal in the Cook Inlet of Alaska (Fig. 4.9) is

Petroleum: Entrapment and Conclusions

an example of an anticline developed in a Tertiary continental sequence which reaches a cumulative thickness of nearly 10 000 m in a basin bounded by more consolidated uplifts (Boss *et al.*, 1976).

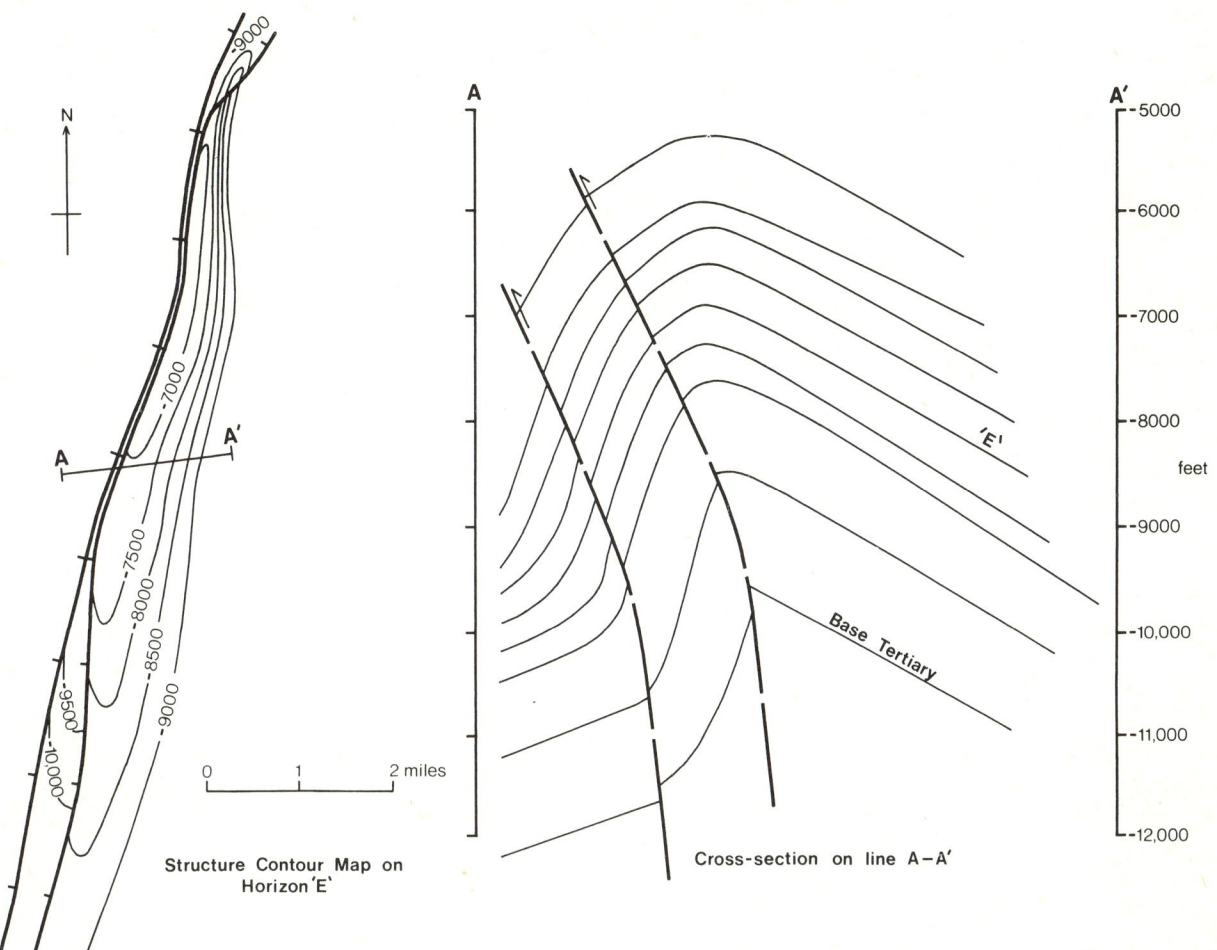

Fig. 4.9. Middle Ground Shoal, Cook Inlet, Alaska. A compressive fold in a thick sequence of continental deposits, approaches similar geometry (after Boss *et al.*, 1976).

The exploration and geometrical interpretation in depth of these structures may not be easy, and will depend not only on the use of all available data on individual and neighbouring folds, but also on an understanding of the tectonic processes involved. For example, although the examples quoted seem clearly due to crustal compression, it is still uncertain in some cases whether or not gravitational décollement gliding may have been involved. Since crustal compression is primarily an expression of destructive plate boundaries, an application of plate tectonic principles should help us in interpreting individual folds.

Chapter 4

NON-COMPRESSIONAL ANTICLINES

It must be appreciated that petroleum trap-forming anticlines can be developed in situations which were not due to compressional folding, and that their geometries will be very different. On a regional scale a basement upwarp in a stable shelf area can produce a broad anticlinal arch in the overlying sediments, and the example of the Williston Basin has already been cited (p. 64 and Fig. 3.12). More locally, draping or compaction over a basement uplift, possibly block-faulted, can form an anticline, and a similar effect can be produced by a carbonate reef or, on a small scale, by an isolated sand body. Such anticlines will increase in amplitude downwards, where they may also be affected by normal flank-faulting which is masked at higher levels. Clearly basement and/or stratigraphical trends will have a controlling influence on the location, size and shape of such structures: it is believed, for example, that some of the giant oilfields of Saudi Arabia are in structures of this type, related to vertical movements on long-established trends (Fig. 3.14).

Other anticlines may be formed by halokinetic movements, either directly above salt domes or plugs, or as residual 'turtle-back' anticlines between the major domes (Fig. 4.10). Such structures are, of course,

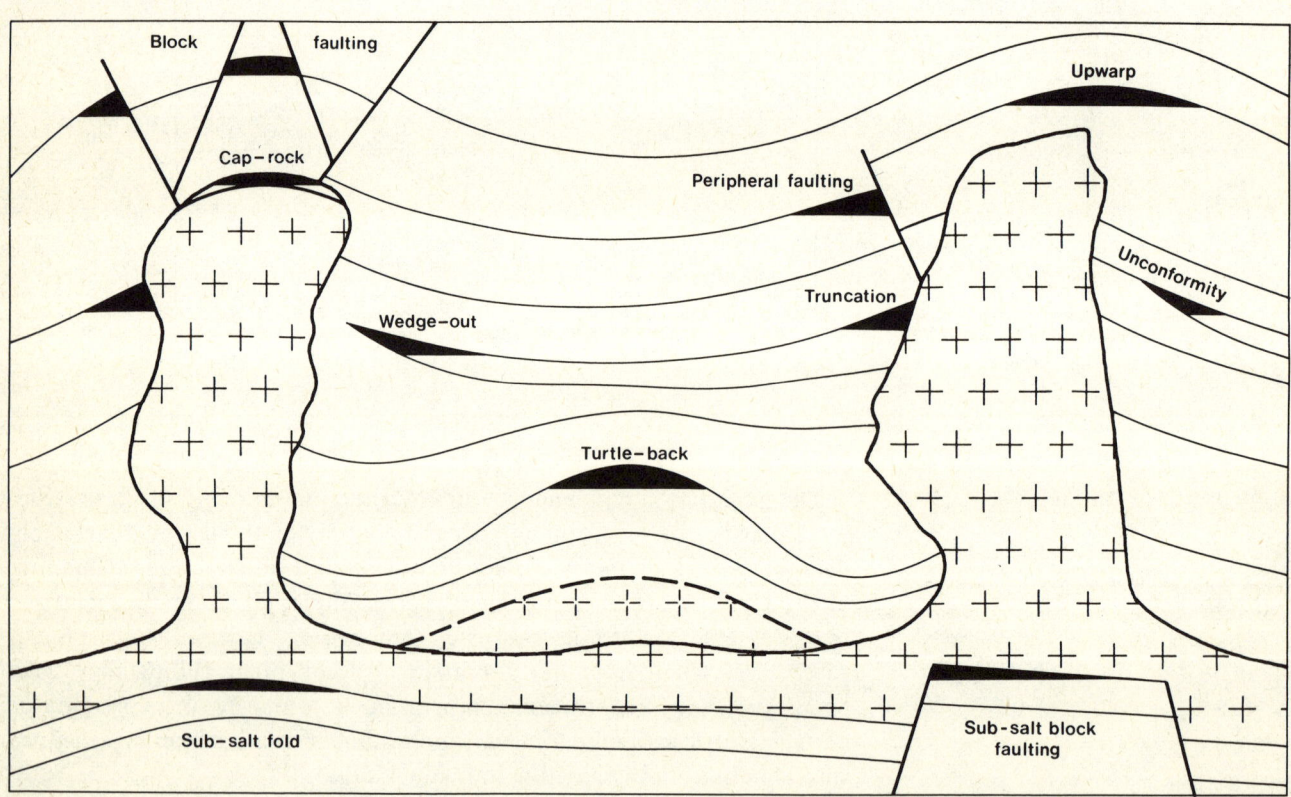

Fig. 4.10. The variety of petroleum traps associated with salt structures. Note that the turtle-backs between the major plugs result from the development of rim-synclines, and may or may not contain a bulge of residual salt.

associated with considerable faulting, which also forms a myriad of small traps, and they can be found in most evaporite-bearing basins. The literature on salt structures and their petroleum occurrences is vast (e.g. Halbouty, 1967); we will content ourselves with noting their presence in northern Europe and the southern North Sea (e.g. Christian, 1969). Still other non-compressional anticlines are the roll-over structures associated with synsedimentary growth faults and especially characteristic of thick, rapidly accumulated regressive sequences, particularly off deltas; reference was made to these in Chapter 3 (p. 69 and Fig. 3.17).

As a final group of anticlines, there are those formed as second-order effects of strike-slip faulting. Structures are commonly diverse and complex, with considerable associated subsidiary faulting. By reference to the Los Angeles Basin, Harding & Lowell (1979) have illustrated the importance of short *en échelon* anticlines developed along the lines of both well-developed and incipient master faults.

FAULT TRAPS

Petroleum is trapped by faults, both normal and reversed, associated with all the types of anticline that we have reviewed and it is not necessary to consider their settings again individually. In most cases, the accumulations can occur on either side of the fault, but there are also traps which, although they might be expected to contain hydrocarbons, are barren. Clearly the critical factor is whether the fault itself can act as a seal, or whether it might provide a channel for further migration. This is a subject which is little understood but fortunately, and provided that the reservoir is juxtaposed against an impermeable formation, nature is commonly kind to us! Weber & Daukoru (1975) studied the problem in relation to growth faults in the Niger Delta which, by virtue of their geometry, have pull-apart tendencies. They found that the major growth faults appear to act as channels of migration, but that single faults associated with roll-over anticlines 'are nearly always sealing as the upthrown side': the sealing qualities of the secondary growth faults are variable, being due to shale smeared into the fault zone from the downthrown side in proportion to the percentage of shale in the interval that has passed a particular upthrown horizon.

The subject of fault traps cannot be left, however, without again stressing the importance of basement block-faulting, both in rift and intracontinental settings. Especially where a continental or shallow marine sequence is up-faulted so that reservoirs may be juxtaposed with overlying source rocks, important petroleum accumulations can be present. Not only do such situations provide extensive trapping conditions, but they may also, as already demonstrated with examples, influence the pattern of subsequent sedimentation and unconformities thereby continuing to affect the distribution of hydrocarbons in younger sediments. Again the North Sea may be quoted where, for example, the Brent Field (Bowen, 1975) provides us with an illustration of an oil-sand elevated *vis-à-vis* its source rock by block-faulting, and actually sealed by a younger sequence above an unconformity (Fig. 3.4): it could be regarded in this sense as a combination trap.

We are now in a position to summarise the variety of traps that we can expect to predominate in different plate tectonic settings. The word 'predominate' is used since nature is almost infinitely varied and what follows must not be regarded as exclusive: petroleum geologists and geophysicists must continually be on the watch for the unusual and the unexpected!

Rift basins
Structures related to tensional faulting are characteristic. Traps in the earlier stages of basin formation are likely to be fault controlled, with the possibility of facies variations in a predominantly continental sequence. Marine transgression can produce truncation traps in the underlying beds, as well as its own stratigraphic traps through areally restricting facies—often in carbonates. Evaporites, if present, will lead to a wide range of both structural and stratigraphic traps in the overlying, usually marine, beds; and basement movements can still be responsible for fault and drape structures, localised unconformities and facies changes.

Fore-arc basins
Traps related to oceanwards thrusting will predominate, although in such tectonically active regions rapid changes of facies, commonly deep water, can be anticipated with the possibility of stratigraphic traps; these, however, are likely to be extremely difficult to locate! In successor basins instability-induced normal faulting can lead to very different trapping situations.

Inter-arc basins
Structures here appear to be both tensional and compressional, so that both anticlinal and fault traps can be anticipated. Again, since subsidence and sedimentation are rapid, stratigraphic controls become important. Conditions in other *intermontane* basins may be similar.

Back-arc basins
Steep dipping thrusts directed backwards from the ocean dominate the margins adjacent to the volcanic arc and could create structural, and perhaps stratigraphic, traps although, for other reasons, such margins

Petroleum: Entrapment and Conclusions

may be poorly prospective. In the central and inner parts of back-arc basins, on the other hand, block movements of the basement can produce fault, drape/compaction anticlinal, unconformity and restricted facies types of trap. In a continental setting sediments derived from the craton could also give rise to stratigraphic traps, although they might be expected preferentially to be open up-dip.

Transform fault basins.
Such basins are complex and rapidly evolved, so that a wide range of anticlinal, fault-closed and stratigraphic traps may be present. To judge from the Californian examples, the largest accumulations occur in anticlines directly related to the strike-slip faults.

Intracratonic basins
Trap developments are similar to those of the inner parts of back-arc basins and of the continental shelves at inactive margins. There may, however, be a tendency for them to be concentrated over regions of basement upwarping, as observed in the Williston Basin (Fig. 3.12).

Inactive continental margins
Traps inherited from the earlier stages of 'rift-drift' basin development will characterise the more deeply buried parts of the sequence. However, once relative stability has been reached and transgression perhaps become more widespread, trap development will continue through basement and halokinetic movements and through the displacement of sedimentary facies. Anticlines will be non-compressional and often very broad and gentle; fault traps are perhaps less common, as basement displacement is absorbed in the sedimentary cover; traps formed by unconformity truncation may be rare, but extensive onlap and facies-related traps can be anticipated.

Localised depressions of the sea-floor may be especially responsible for the distribution of the stratigraphic traps, which can also develop from a general oceanwards progradation. In particular this will be the case off major deltas, where additional traps in roll-over anticlines and fault-sealed reservoirs may be associated with growth-faulting.

Fore-deep basins
A basin developed marginally to, and perhaps partly within, a fold-belt resulting from the collision of an inactive continental margin with an opposing continent, will display a twofold range of traps. It will still contain those derived from its inactive shelf history, but these may be subordinated to anticlinal and thrust-fault traps formed by the orogeny.

Chapter 4

The latter can be expected to dominate the early stages of an exploration programme, but the more obscure traps may become important later.

From this brief analysis it will be clear that plate tectonic control on petroleum traps is fundamental. Consideration of the principles involved can help immeasurably in predicting, understanding and interpreting their nature and distribution.

4.3 Cap-rock or seal†

Oil segregates upward in a reservoir rock in much the same way, and for much the same reason, that cream rises to the top of the milk. The function of the cap-rock or seal is to restrict, or actually prevent, the buoyant rise of the hydrocarbons through the water-saturated rock column. In the commonest instances it does this by presenting an array of migration pathways, which are of substantially smaller diameter, to the secondarily migrating oil globules and gas bubbles.

Under these circumstances, and provided that no hydrodynamic drive occurs, the *capillary pressure*, necessary to distort the migrating droplets and force them through the pore throats, commonly exceeds the buoyancy forces acting on the droplets and thus upward migration will cease. However, the arrival of more hydrocarbon droplets at the reservoir-seal interface, and their coalescence with their forerunners, sees a gradual increase in the height of the oil column and a resultant increase in the buoyant force. A 3-m oil column, for example, would exert a buoyant force sufficient to cause migration through most water-saturated reservoir rocks (Schowalter, 1979). Yet many accumulations feature oil columns five or ten times as great. Evidently, therefore, a seal must be characterised by pore throats very significantly smaller than those of the typical reservoir rocks. This is merely to regularise the observation that sedimentary rocks of very fine grain, such as claystones, or of effectively zero porosity, such as evaporites, constitute the best seals.

Since c. 50% of all sedimentary rocks are mudrocks, with the requisite fine grain, there is no great incentive to research the provenance and conditions of deposition of the typical seal facies—nature provides them in abundance. But certain mildly intriguing problems exist. For example, do clays from tropical weathering zones, richer in kaolinite and gibbsite, constitute better potential seals than those from the present polar latitudes, which have higher relative proportions of chlorite and illite? Or, given the more or less worldwide tropicality of the Mesozoic climate, do clays from a Mesozoic cratonic interior constitute a better potential seal than those, for example, from a contemporary, rapidly eroding, volcanic

† Sections marked thus are written by R.J. Bailey and published by permission of The British Petroleum Company Ltd.

Petroleum: Entrapment and Conclusions

arc? If affirmative evidence is found in either case, then the movements of continents relative to the climatic zones and the changing patterns of lithosphere subduction might be of some (slight) relevance to the effectiveness of a clay/shale/mudrock seal. Only in such circumstances could these ubiquitous seals be said to relate to plate tectonics.

By contrast, the palaeoclimatic and, to a lesser degree, tectonic controls on the distribution of evaporites provide a more obvious relationship. They have a well-known association with the early phases of crustal rifting (Chapter 3), with intraplate desert basins of the continental interiors, and with back-reef environments in continental interior and continental margin intraplate basins—all of these at low palaeolatitude.

Perhaps the most intriguing form of seal, however, is that which also constitutes the hydrocarbon source. In the Rengiu oilfield of the North China (Tertiary) Basin, for example, 'buried hills' of fractured and weathered Precambrian limestones constitute a prolific reservoir sourced and sealed by Palaeogene lacustrine shales (Yan & Zhai, 1980). A marine shale seal with the additional attribute of source potential, such as is seemingly encountered in the analogous Amposta oilfield of offshore South Spain (Siñeriz *et al.*, 1978) of course partakes of the plate tectonic relationships of all marine source rocks (section 2.2).

4.4 The deep-water environment†

In the matter of seal, reservoir, source and structure, the connection between the modern strategies of petroleum exploration and the tenets of plate tectonics is often difficult to establish. In many instances the theory has provided the conceptual framework for modern exploration without, in fact, generating any more exploration success than would otherwise have been achieved. The reason for the dubious benefits of plate tectonic theory in this respect is not difficult to appreciate. Most petroleum accumulations reside in epicontinental or intracratonic basins where a pragmatic exploration policy—'drill the big structures'—can yield results regardless of the origin of the structures. However, it is possible to point to one growing sphere of petroleum exploration which, from the outset, has depended, to a degree, on the direct application of plate tectonic theory. This is the deep-water exploration for oil and gas.

In the early 1970s petroleum exploration was poised at the outer limits of the continental shelves. Its progress to this new exploration frontier had involved the extension of well-tried 'continental' exploration methods to the drowned continental borders. Exploration teams had got their feet wet, but the world ocean represented a geological realm quite outside their experience. Admittedly, marine seismic reflection profiling (see Vail *et al.*, 1977) had already provided a powerful means to gain this experience, but the essentials of the problem came to be understood without recourse to costly programmes of seismic shooting.

In essence, the mounting evidence of the global applicability of plate tectonic theory allowed much more confident use of the limited data to hand in predicting the geology of unexplored oceans and continental margins. From the early results of *Glomar Challenger* drilling, for example, came the realisation that the ocean ridges and the oceanic crust for thousands of kilometres on either hand probably support a sedimentary sequence too thin and too pelagic in character to warrant any exploration interest. Also, plate tectonic theory asserted the otherwise obscure kinship of the modern Red Sea and the Mesozoic Atlantic; so that when Lowell & Genik (1972) came to report the results of petroleum exploration of the Red Sea's margins, they did so in terms of a sequence of crustal arching and rifting which became a model for Atlantic-type margins of whatever age. It is clear that without the theory there would have been little or no sanction for the widespread use of the Red Sea model. Its development had, in fact, an important influence on deep-water exploration strategy and it is worth examining in these terms.

The basis of the model, as it has evolved in more recent years, is that there occurs ductile thinning of the deeper continental lithosphere and brittle deformation, with listric normal faults, at higher, crustal and sub-crustal, levels where rifting is manifest. It became clear that the transition from rifting to drifting, involving active sea-floor spreading, was accompanied by an important change in tectonics. So long as a discrete lithosphere plate was being pulled apart, strain accumulated in a rift zone. Grabens, half grabens, horsts and tilted fault blocks developed in continental crust, the subsiding grabens typically filling with 'syn-rifting' sequences of non-marine clastics. But when the original lithosphere plate finally separated into two units, partitioned by a septum of hot asthenosphere along the new ocean ridge, it was this latter feature which became the locus of virtually all the tensional stress. Along the two recently formed continental margins normal faulting effectively ceased and passive subsidence (see below) permitted marine facies to transgress landward, thus creating an unconformity which stepped from fault block to fault block and was rarely thereafter penetrated by the faults. Typically this classic kind of 'break-up' unconformity is buried by a thick and little-deformed marginal sedimentary prism recording post break-up subsidence of the continental trailing edge. The exploration of the deeper continental margin of north-west Australia, for example (Falvey, 1974; Veevers & Cotterill, 1978; Branson, 1978; Exon & Willcox, 1978), is to a large extent concerned with actual and potential hydrocarbon trapping at the break-up unconformity; and 'pre break-up' hydrocarbon plays are now routinely sought in any new deep-water exploration of an Atlantic-type continental margin.

In this same context, a new focus of interest is in the possibility of inferring the thermal history of a marginal sedimentary sequence. Geothermal heating is crucial to the maturation of hydrocarbons (e.g. Tissot & Welte, 1978). In the epicontinental basins a general equation between time and depth of burial, on the one hand, and geothermal heating, on the other, allows us to say that sediments buried to depths of 3 km or so for some tens of millions of years will very probably have achieved thermal maturity with respect to the generation of oil and, possibly, gas. We are still far from appreciating the vagaries of heat-flow through old continental lithosphere, so that, in the absence of actual well data, there has been a tendency to rely on some such rule of thumb. The post break-up subsidence of an Atlantic-type margin, however, can be represented in terms of the conductive cooling and thickening of oceanic lithosphere initially, along its oceanic edge, at melting temperature (Turcotte & Oxburgh, 1967). Thus, when allowance is made for sediment loading and eustatic effects (e.g. Steckler & Watts, 1978), there remains a component of subsidence which can be expressed in terms of lithospheric heat loss.

In fact, equations can be written which specify the temperature history of any 'time slice' through the marginal sedimentary prism (Turcotte & McAdoo, 1979). This holds out the possibility of plotting the theoretical ascent through the sedimentary section of an isotherm representing the *oil generation threshold*, or of determining at what stage an inferred source rock may have achieved the corresponding thermal maturity. Herein perhaps lies the most direct and practical application to oil exploration of the ideas collected under the general heading of plate tectonics.

In the exploration of deep-water marginal environments of non-Atlantic type, confidence in an extrapolation from the particular (one traverse of one slope) to the general (all Andean margins), again stems from the belief in the likelihood that the same *kinds* of tectonics will characterise all margins where subduction occurs. In these instances, however, the data are largely of a negative character, for the thick sedimentary developments which characterise the walls of the oceanic trenches are locally proved and generally expected to comprise a package of successively underthrust, imbricate wedges of *pelagic* sediment (e.g. DSDP Leg 31, site 298, 1973; Beck & Lehner, 1974, 1980). There is little hope that such highly deformed sequences will anywhere prove prospective.

Finally, it is chastening to consider that in a recent analysis of deep-water prospects, in which context plate tectonics arguably has its greatest utility as an aid to petroleum exploration, Hedberg *et al.* (1979) conclude that the small ocean basins are the most promising areas for petroleum generation. The corrective to our enthusiasm for plate tectonics in this context lies in the fact that these small ocean basins constitute probably the least understood class of large-scale features attributable to the workings of plate tectonics and that they are prospective by virtue of their combination of the characteristics which classically typify the prospective basins of the continental interiors: thick sedimentary section; deposition under conditions of restricted circulation; reservoir development in deltaic (and somewhat analogous deep-water fan) facies; and a history including tectonism. In this sense, therefore, exploration has come full circle, returning even in the oceanic realm, to the rules established by experience in the days prior to plate tectonic theory.

4.5 Petroleum: general conclusions*

In these three chapters we have attempted to review the possible impact of plate tectonics on the habitat of petroleum, by considering the various factors affecting its formation and distribution. The extent to which this influence is recognised will still depend on the viewpoint and bias of the individual geologist, and the reader will not have failed to notice a certain difference of view between the two contributors to these chapters. Such divergences of opinion are healthy since they help to ensure a continuing examination of the subject. We therefore invite readers to ponder the possible controls on petroleum occurrence and to formulate their own opinions.

We must, however, reiterate our concern that the problem does not become a matter of mere semantics. Thus, for example, while one geologist may argue that plate tectonics controls the evolution of sedimentary basins and hence all aspects of their geology, another may take the view that since petroleum is endemic to the sedimentary basin its habitat then reflects only local factors. Arguments such as this will arise unless the problem is clearly defined, and they can become unhelpful and pointless. Statements to the effect that 'Plate tectonics expresses the way the Earth evolves and therefore influences all geological considerations', or 'Plate tectonics hasn't found us a barrel of oil and so we can live without it' fall into this unhelpful category, and they serve only to polarise standpoints.

Thus, what we have aimed to do is to consider the various parameters governing the formation and entrapment of petroleum, and how our understanding of them may be guided or influenced by the concepts of plate tectonics. So as to assist readers to make up their own minds, we conclude with a brief statement of the factors that help to make the Middle East the world's most prolific petroleum-bearing basin. Readers may consider for themselves the extent to which those factors are a consequence of plate tectonics.

THE MIDDLE EAST PROBLEM

The possible reasons for the richness of the Middle East were discussed admirably in 1974 by Irving *et al.*, and the reader is referred to their paper for amplification of some of the following points.

Throughout most of the Mesozoic and Early Cenozoic, the site of the Middle East Basin formed a part of a very broad continental shelf at the inactive northern margin of the Afro-Arabian continent. It lay at the southern side of a 'Southern Tethys Ocean' which, after an initial marine invasion in the Permian, had developed as a result of Triassic continental rifting. From the Triassic onwards the margin was inactive until, in the Miocene, the southern Tethys was closed again through collision of the Afro-Arabian shelf with a separate, more northerly continental block: this collision was the cause of the Zagros folding that affected the outer part of the shelf. The Middle East shelf evolved in tropical latitudes. The southern Tethys oceanway was perhaps somewhat restricted, being limited to a few hundreds of kilometres in width and probably closed temporarily at its western end. Oceanic circulation is likely to have been affected by prevailing easterly winds, and to have influenced the abundance of organic matter over the marginal shelf seas.

The petroleum source rocks were mainly the fine-grained, predominantly argillaceous and organic-rich Mesozoic sediments that accumulated in localised and intermittent depressions within the shelf: the underlying Palaeozoic also contributed over parts of the region. Except in these depressions, the shelf deposits were principally shallow-water carbonates with the episodic influx of quartz sand from the Arabian Shield. Reservoir rocks are thus in carbonates, usually the coarser facies modified by diagenesis, and in sheet sands. Fluctuations of sea-level led to numerous interruptions in sedimentation, and also to widespread evaporites which are particularly prominent near the Jurassic-Cretaceous boundary and in the Miocene. Cap-rocks are provided by these evaporites, as well as by the argillaceous intervals and impermeable phases of the carbonates. For palaeogeographical reasons, therefore, the region is unusually well endowed with source, reservoir and cap-rocks, in various parts of the stratigraphical column.

Traps are related to three causes:

1 differential vertical movements within the Shield basement, at least partially associated with block-faulting, and draping, facies changes and compaction within the sedimentary overburden;

2 diapiric movements of an underlying layer of Infra-Cambrian salt;

Chapter 4

3 in the outer part of the shelf only, compressional folding due to the Miocene collision of the shelf with a central Iranian continent, resulting in a localisation of Mesozoic petroleum into Neogene traps. In the latter region especially, extensive fracturing aids reservoir performance.

It is the abnormally favourable coincidence of these various factors that has provided the vast petroleum resources of the Middle East. *Has plate tectonics had anything to do with it?*

References

Adasani, M. (1965) The Greater Burgan Field. *Fifth Arab Petroleum Congress Cairo.* Paper 35 (B-3) 27 pp.

Beck, R.H. & Lehner, P. (1974) Oceans, new frontier in exploration. *Bull. Amer. Assoc. Petrol. Geol.*, **58**, 376-395.

Boss, R.F., Lennon, R.B. & Wilson, B.W. (1976) Middle Ground Shoal Oil Field, Alaska. In *North American Oil and Gas Fields* (Ed. J. Braunstein) Am. Assoc. Petrol. Geol., Mem. 24, pp. 1-22.

Bowen, J.M. (1975) The Brent Oil Field. In *Petroleum and the Continental Shelf of Northwest Europe, Vol. 1. Geology* (Ed. A.W. Woodland) Applied Science Publishers, London, pp. 353-362.

Branson, J.C. (1978) Evolution of sedimentary basins from Mesozoic times in Australia's continental shelf and slope. *Tectonophys.*, **48**, 389-412.

Christian, H.E. (1969) Some observations on the initiation of salt structures of the southern British North Sea. In *The Exploration for Petroleum in Europe and North Africa* (Ed. P. Hepple) Inst. Petrol., London, pp. 231-250.

Colmann-Sadd, S.P. (1978) Fold development in Zagros simply folded belt, southwest Iran. *Bull. Amer. Assoc. Petrol. Geol.*, **62**, 984-1003.

Dickinson, W.R. & Suczek, C.A. (1979) Plate tectonics and sandstone compositions. *Bull. Amer. Assoc. Petrol. Geol.*, **63**, 2164-82.

Evamy, D.D., Haremboure, J., Kamerling, P., Knaap, W.A., Molloy, F.A. & Rowlands, P.H. (1978) Hydrocarbon habitat of Tertiary Niger Delta. *Bull. Amer. Assoc. Petrol. Geol.*, **62**, 1-39.

Exon, N.F. & Willcox, J.B. (1978) Geology and petroleum potential of Exmouth Plateau area off Western Australia. *Bull. Amer. Assoc. Petrol. Geol.*, **62**, 40-72.

Falcon, N.L. (1969) Problems of the relationship between surface structure and deep displacements illustrated by the Zagros Range. In *Time and Place in Orogeny* (Eds P.E. Kent *et al.*) Geol. Soc. Lond. Spec. Publ. 3, pp. 9-22.

Falvey, D.A. (1974) Development of continental margins in plate tectonic theory. *Austral. Petrol. Expln. Assoc.*, **14**, 95-106.

Glennie, K.W. (1972) Permian Rotliegendes of north-west Europe interpreted in light of modern desert sedimentation studies. *Bull. Amer. Assoc. Petrol. Geol.*, **56**, 1048-71.

Halbouty, M.T. (1967) *Salt Domes: Gulf Region, United States and Mexico.* Gulf. Publ. Co., Houston, 425 pp.

Harding, T.P. & Lowell, J.D. (1979) Structural styles, their plate-tectonic habitats, and hydrocarbon traps in petroleum provinces. *Bull. Amer. Assoc. Petrol. Geol.*, **63**, 1016-1058.

Hedberg, H.D., Moody, J.D. & Hedberg, R.M. (1979) Petroleum prospects of the deep offshore. *Bull. Amer. Assoc. Petrol. Geol.*, **63**, 286-300.

Irving, E., North, F.K. & Couillard, R. (1974) Oil, Climate and Tectonics. *Can. J. Earth. Sci.*, **11**, 1-17.

Lees, G.M. (1952) Foreland folding. *Q. J. geol. Soc. Lond.*, **108**, 1-34.

Lowell, J.D. & Genik, G.J. (1972) Seafloor spreading and structural evolution of Southern Red Sea. *Bull. Amer. Assoc. Petrol. Geol.,* **56**, 247-259.

Moody, J.D. (1975) Distribution and geological characteristics of giant oil fields. In *Petroleum and Global Tectonics* (Eds A.G. Fischer & S. Judson) Princeton University Press, Princeton, pp. 307-320.

Rittenhouse, G. (1972) Stratigraphic trap classification. In *Stratigraphic Oil and Gas Fields—Classification, Exploration Methods and Case Histories* (Ed. R.E. King) Am. Assoc. Petrol. Geol., Mem. 16, pp. 14-28.

Schowalter, T.T. (1979) Mechanics of secondary hydrocarbon migration and entrapment. *Bull. Amer. Assoc. Petrol. Geol.,* **63**, 723-760.

Schlumberger (1975) *Well Evaluation Conference—Arabia.* Schlumberger, France, 152 pp.

Selley, R.C. (1978) *Ancient Sedimentary Environments.* 2nd Edition. Chapman & Hall, London, 287 pp.

Sinerz, B.G., Querol, R., Castillo, F. & Arribas, J.R.F. (1979) A new hydrocarbon province in the W. Mediterranean. Preprint, 10th World Petroleum Congress, Bucharest 9-14 September 1979.

Steckler, M.S. & Watts, A.B. (1978) Subsidence of the Atlantic type continental margin off New York. *Earth Planet. Sci. Lett.,* **41**, 1-13.

Talwani, M. and Pitman, W.C. III (Editors) (1977) *Island arcs deep sea trenches and back-arc basins.* Amer. Geophys. Union Maurice Ewing Series 1.

Tissot, B.P. & Welte, D.H. (1978) *Petroleum formation and occurrence.* Springer Verlag, Berlin, 538 pp.

Turcotte, D.L. & Oxburgh, E.R. (1967) Finite amplitude convection cells and continental drift. *J. Fluid Mech.,* **28**, 29-42.

Turcotte, D.L. & McAdoo, D.C. (1979) Thermal subsidence and petroleum generation in the SW block of the Los Angeles basin, Calif. *Jour. Geophys. Res.,* **84**, 3460-3464.

Vail, P.R., Mitchum, R.M., Jr & Thompson, S., III (1977) Seismic stratigraphy and global changes of sea level. In *Seismic stratigraphy applications to hydrocarbon exploration.* (Ed. C.E. Payton) Amer. Assoc. Petrol. Geol., Mem. 26, pp. 49-212.

Veevers, J.J. & Cotterill, D. (1978) Western margin of Australia: evolution of a rifted arch system. *Bull. Geol. Soc. Amer.,* **89**, 337-355.

Weber, K.J. (1971) Sedimentological aspects of oil fields in the Niger Delta. *Geol. en Mijnbouw,* **50**, 559-576.

Weber, K.J. & Daukoru, E. (1975) Petroleum Geology of the Niger Delta. *Proc. 9th World Petrol. Congr. Tokyo* Vol. 2, pp. 209-221.

Williams, J.J. (1972) Augila Field, Libya: depositional environment of sedimentary reservoir and description of igneous reservoir. In *Stratigraphic Oil and Gas Fields—Classification, Exploration Methods and Case Histories* (Ed. R.E. King) Amer. Assoc. Petrol. Geol., Mem. 16, pp. 623-632.

Yan, D. & Zhai, G. (1980) Exploration practice and Prospects in respect to the buried-hill oilfields in N. China. In *UN International Meeting on Petroleum Geology, Beijing, China.*

Chapter 5

Coal

G.A.L. JOHNSON

5.1 Introduction

Coal is a combustible rock formed by partial decomposition and diagenesis of accumulations of terrestrial and freshwater vegetation. It consists of a mixed assemblage of plant debris which varies with the nature of the plant fragments, the conditions of deposition and the extent of diagenesis. The resultant diversity of mineralogy and chemical composition gives us the different coal types. Plants have contributed the essential ingredients for the formation of coal since their first appearance in the geological record. Recognisable land plants are known in some abundance from the Silurian, but the first fossil peat deposits and coal seams are Devonian. By Carboniferous times, about 360 million years ago, the land flora included representatives of several of the major plant groups capable of producing both forest trees and ground cover vegetation. Luxuriant swamp-forests developed and thick peat deposits were laid down which are now preserved within Carboniferous strata as coal seams. Problems of reproduction and migration in these plants had been solved by the production of vast quantities of fertile spores and they colonised the equable climatic zones of the world to provide an almost cosmopolitan flora. During the Carboniferous floral divergence took place, owing to plant evolution with climatic and geographical isolation, and by the Permian five floral provinces were established (Chaloner & Lacey, 1973). From these floras the cycadioid plants of the Mesozoic evolved and they formed the parent materials for the Triassic, Jurassic and Cretaceous coals. During the Cretaceous the flowering plants (Angiosperms) became established and at the beginning of the Tertiary they rapidly became dominant in the land flora. Coal derived from the Tertiary flowering plants makes up more than half of the total known world reserves.

During life plants synthesise their tissues from atmospheric carbon dioxide and soil water. Organic compounds that make up the plant tissues include dominant carbohydrates, such as cellulose, starches and sugars, and less abundant resins and oils. The latter are hydrocarbon-rich materials as compared with carbohydrates. In living plants they serve a protective function and are resistant to decay. Organic deposits formed of hydrocarbon-rich plant fragments will differ greatly in chemical composition from unsorted plant debris. The type of coal formed much depends on the nature of the plant primary materials and the varying proportions of carbohydrate and hydrocarbon constituents.

The first process in the formation of coal from plant material is biochemical decomposition. Under normal sub-aerial conditions plants disintegrate completely under the attack of fungi and bacteria. Decay is stopped when the activity of these organisms ceases often owing to toxic concentrations of waste products. For this to take place deposition in very wet conditions or under water is required. Two processes of biochemical decay are recognised, peat-formation and putrefaction. For peat-formation plant debris accumulates under stagnant water at the site of growth. Plant growth at the surface continually adds to the decaying pile of vegetable material that gradually becomes compressed and altered into brown fibrous peat. Putrefaction takes place under deeper water of open, but stagnant, ponds or lakes. Wind-blown plant fragments including pollen and spores collect on the bottom, with algal remains and inorganic mud, to form a slimy deposit called sapropel. Biochemical putrefaction of the less resistant plant fragments takes place with the formation of marsh-gas (methane) and the deposit, initially hydrocarbon-rich with the concentration of spores, pollen and algae, becomes more enriched with these materials. When water is driven from the sapropel by compression it becomes black, homogeneous and hard.

The conditions under which peats and sapropels form are controlled by climate and topography. Warm-wet climate is required for luxuriant growth of vegetation and constant conditions of stagnant water for the preservation of peat or sapropel. Under conditions of high rainfall, peat-formation takes place over uplands over many parts of the world. The widespread blanket peat deposits over the British uplands is an example. Blanket peat is a most transient deposit and tends to become unstable and erode with slight changes of environmental conditions. It would not, except under exceptional circumstances, have any chance of preservation as coal in a stratigraphical sequence. The environment in which coal-forming peat accumulates is more restricted and requires flat, low-lying ground transformed into swamps by largely stagnant surface water. Broad, poorly drained, paralic coastal areas and flat limnic inland lake basins are ideal for plant growth and peat accumulation in swamp conditions. Slow regional sinking of these regions is essential for preservation of peat layers by burial within a sedimentary succession. Where regional depression and the rate of peat accumulation are well balanced, thick peat deposits are formed with only minor intercalations of barren clastic sediment. Thus in the Gippsland brown coal basin in Victoria,

Coal

Australia, there are up to 300 m of coal divided by only few shale and mud bands; similar basins are known in India and Canada.

Exact balance between subsidence and peat accumulation seldom takes place and the formation of peat or sapropel normally terminates owing to changes in water level. A drop in water level over the swamp region causes changes in vegetation cover and may lead to removal by erosion of the entire peat deposit. A rise in water level, causing flooding of the swamp region, again effects the vegetation and may lead to significant deposition of clastic sediment muds and sands over the peat. Where rapid flooding of low-lying regions of a swamp takes place quick deposition of sediment traps vertical trunks of forest trees and these may be preserved as fossils in their position of growth; vertical tree trunks are not uncommon fossils in coal bearing strata. Deposition of clastic sediment rapidly builds up to water level and the emerging land surface quickly re-acquires vegetation cover allowing the processes of peat-formation to continue. Coal swamp sedimentation normally produces a succession of peat beds separated by bands of shale and sandstone. Where swamps are invaded by marine or limnic waters, limestones, marls and shales are laid down over the peat. A general succession of strata is laid down when paralic or limnic coal swamps are submerged and covered by clastic sediment. First the marine or lake-bottom limestone and shale is deposited, then shale, followed by sandstone and terminating with a mineral soil called seatearth, underlying the peat. This upward coarsening cycle of sedimentation or cyclothem behaves like a delta with prodelta muds followed by deltaic sands prograding outwards from the shoreline.

5.2 The process of coalification

Biochemical changes in peat and sapropel come to an end when the activity of organisms ceases. By this time the formation of peat and sapropel has taken place and the further alteration of these products to coal is brought about by chemical and physical changes, caused by rising pressure and temperature, after burial under younger sediments. Peat and sapropel are compressed and considerably reduced in volume by the expulsion of water and volatile compounds. The degree of compaction from peat to coal probably varies with different deposits but a figure of 10 : 1 is often quoted. With compaction, progressive enrichment with carbon takes place and the peat gradually develops the colour, solidity and composition of coal. In the same way sapropels compact to a tough, dark homogeneous solid after burial.

With compaction peat changes to humic coal which is layered with bright and dull bands. The bright bands are called clarain and are mainly composed of fine laminae of brilliant black vitrinite, the partly decomposed

Chapter 5

remains of bark and wood. The dull layers, called durain, are harder than clarain, lack the brilliant bands of vitrinite and are more homogeneous in appearance. The principal components of durain are micrinite (fine plant debris), spores, resin bodies and cuticles, and fusinite (black mineral charcoal); vitrinite is only present in subordinate amounts. Coals derived from sapropel are like durain in that they are composed of finely disintegrated plant debris, but differ in being made up of vitrinite fragments. Cannel, a sapropelic coal, is predominantly made up of vitrinite together with important amounts of resin, cuticle and spores. Boghead coals and torbanite resemble cannel in appearance and mode of formation. They differ in being hydrocarbon rich owing to substantial quantities of algal remains in their composition. The sapropelic coals are more limited vertically and laterally than the humic coals. The former originate in pools within the swamp while the latter form from the swamp peat itself. In the following pages it is the extensive and economically important humic coals and their derivatives that will be considered rather than the minor sapropelic coals that are associated with them.

The progressive alteration of peat during burial by younger sediments over long periods of time results in coals that vary in properties and chemical composition (Fig. 5.1). Brown coal and lignite are an early stage

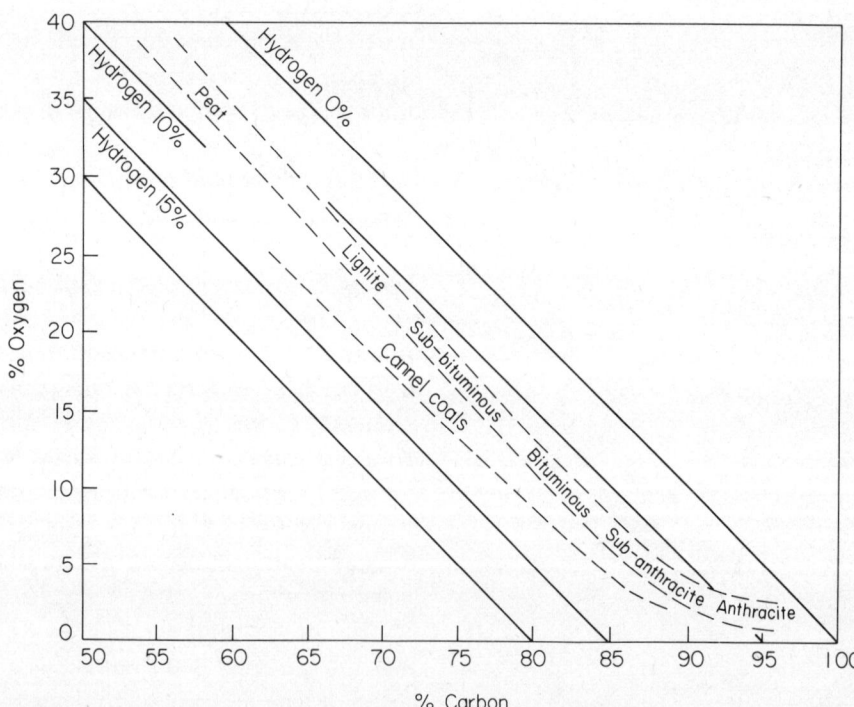

Fig. 5.1. Change in chemical composition from peat, through lignite, sub-bituminous coal, bituminous coal to semi-anthracite and anthracite in ascending rank. Through most of this change hydrogen is almost constant, but it decreases with the formation of anthracite. (After Raistrick & Marshall, 1939.)

in the coalification of peat. Water content is high, volatiles are high and carbon is only slightly increased relative to peat. Lignite is typically more consolidated than brown coal, but both are brown to black coloured and lack lustre. Being only slightly altered they are almost confined to the Recent and Tertiary and are virtually absent in beds older than the Mesozoic. Further alteration causes changes from lignite to sub-bituminous coal and then to bituminous coal, in which clarain and durain are the characteristic constituents.

Sub-bituminous coal is so named because it burns with a smoky flame and tends to melt when heated in the absence of air. It includes house-coal and gas-coal and is the most widely distributed and abundant of the solid fuels. These coals show progressive enrichment in carbon and reduction in volatiles as they increase in rank. Low-rank bituminous coals have 30-40% volatiles and are most suitable for coal-gas production and general use. High-rank bituminous coals, with volatiles between 20 and 30%, can provide a high-grade coke suitable for industrial use. The ability of vitrinite to become plastic on heating and form coke only takes place over a limited range of rank, roughly between 82-89% carbon. The coking coals are particularly important economically. High-rank coals with volatiles between 10 and 20% do not coke freely and provide the best quality furnace or steam coals. A low ash content is desirable in industrial and domestic coals to reduce slag and protect furnace linings. Clarain rich bituminous coals containing little inorganic mineral matter are the most satisfactory fuels.

The highest rank coals, the anthracites, are produced from bituminous coal by thermal and/or dynamic metamorphism. The change to anthracite is gradual, banding in the coal becomes less distinct, it becomes more homogeneous, harder and develops conchoidal fracture. Chemical changes associated with this liberate volatiles, but instead of carbon dioxide and water that are produced as bituminous coals increase in rank, the anthracites produce methane and water (Fig. 5.1). In the final stages of increasing rank the hydrogen content is reduced more rapidly than oxygen to produce anthracite. Methane gas in economic quantities may be evolved during this diagenetic change towards anthracite. (The natural gas of the southern North Sea Basin has its origin in change to higher rank in deeply buried Upper Carboniferous coals.)

Variation in the rank of coals broadly follows Hilt's Law which states that the rank of a coal is a function of the depth to which it has been buried. In any coalfield the upper coals are of lower rank than the lower coals. Although this is broadly correct there are many exceptions. Temperature metamorphism adjacent to dykes produce an aureole of high-rank coal that can be economically important. The Ludworth Dyke of

Chapter 5

Late Carboniferous age produces low-volatile, heat-altered coals for more than one kilometre on either side in the Easington area, Co. Durham (Smith & Francis, 1967, p.258). Regions of abnormal high heat-flow from the interior of the earth can produce significant rank changes in the overlying coal seams. This effect has been detected in Northern England over the sub-surface Weardale granite, but thermal effects from sill intrusions are also important here (Creaney, 1980). Dynamic metamorphism associated with earth movements also causes rank changes in coal. Frictional heat produced in major thrusts is known to have inverted rank changes in coal seams above and below a thrust. High-rank anthracites occur in highly folded orogenic belts associated with the late Carboniferous Hercynian earth movements in western Europe and North America; the south Wales coalfield in Britain and the Pennsylvania anthracite field of USA are examples.

5.3 The depositional environment of coal

The essential requirements for peat-formation that can lead to coal deposits are, tropical or warm temperate climate for continuous and luxuriant plant growth in a setting of sinking, waterlogged ground flooded with shallow still water in which peat accumulation can take place. Geologically these conditions are found in lowland areas of deposition in contrast to uplifted regions of erosion. With exceptions in Brazil and east Siberia, the ancient stable Precambrian shield areas do not provide the conditions necessary for coal formation. Peat accumulation at the surface in these stable regions is not preserved by burial by younger sediments owing to lack of subsidence and eventually it becomes unstable and is removed by erosion. The less stable platform regions skirting the shield areas give conditions of downwarping in which peat accumulation, burial and coal formation take place. These processes occur within continental interior lake basins or limnic basins and on wide shallow shelf regions developed on some continental margins—the paralic environment. If swamp peat is to be preserved and converted to coal by burial under younger strata, both limnic and paralic environments must have a structural setting of regional sinking. This combination of conditions is most often found in post-orogenic intermontane limnic basins within continents and on aseismic, Atlantic-type paralic marine margins to continents.

LIMNIC COAL BASINS

Limnic coal-forming conditions are found in sinking basins often in the vicinity of newly uplifted mountain ranges. Sedimentation is terrestrial and completely cut off from the sea so that no marine intercalations are found in the deposits. Deposition is in sinking troughs adjacent to rising mountain ranges and are 'yoked' basins within the continental interior. They are usually discrete, circular or elongate regions of continental deposition. Sedimentation varies from alluvial fan and alluvial apron on the arid and semi-arid margins to flood plain, ephemeral and perennial lake sediments, lacustrine deltas and peat swamps in the central region. The sediments are often red-coloured, coarse conglomerates and arkosic sandstones interbedded with grey siltstones and shales. The tectonic environment of post-orogenic intermontane basins in which these sediments are laid down occurs after continental collision orogeny and also after Andean type orogeny at active continental margins.

Carboniferous and Permian age limnic coal basins are widely developed in central Europe (Fig. 5.2). The major Hercynian orogeny during the Namurian in this region is believed to have been of the Andean type and produced elevated and extensive new mountain ranges. This active continental margin formed on the southern side of a marginal basin that crossed Europe in an east to west direction. The basin tended to close during the Carboniferous and developed a southward dipping zone of subduction to form the Andean or Pacific margin (Anderson, 1975; Johnson, 1978). Intermontane basins between the newly uplifted mountain ranges were established during the Upper Carboniferous and limnic coal deposition took place during Westphalian, Stephanian and into the Permian.

In northern France limnic coal basins are known at Saar and Lorraine (Fig. 5.2) where 6000 m of Westphalian and Stephanian continental strata contain 88 coal seams with exploitable seams averaging 3-4 m thick (Gignaux, 1955). Similarly in Brittany, the Vosges and the Massif Central, many small limnic coalfields, mainly of Stephanian age, have developed in Hercynian intermontane basins. The principal phase of Hercynian folding was between the Dinantian and the productive Coal Measures which in some localities started in the Early Westphalian. Further to the east in Bohemia, Czechoslovakia and Poland, another series of eight limnic coal basins, some of considerable area, formed under the same conditions as those in France (Fig. 5.2). They occupy intermontane depressions between crystalline massifs uplifted during the same orogenic episode between Dinantian and Westphalian. Deposition of continental sediments within the depressions persisted from Westphalian into the Permian without break and continued, in two basins, into the Triassic (Holub, 1977). Continental sequences more than 1000 m thick are present in the basins with thick coal seams developed widely, such as the Upper Radnice seam averaging 6-8 m thick.

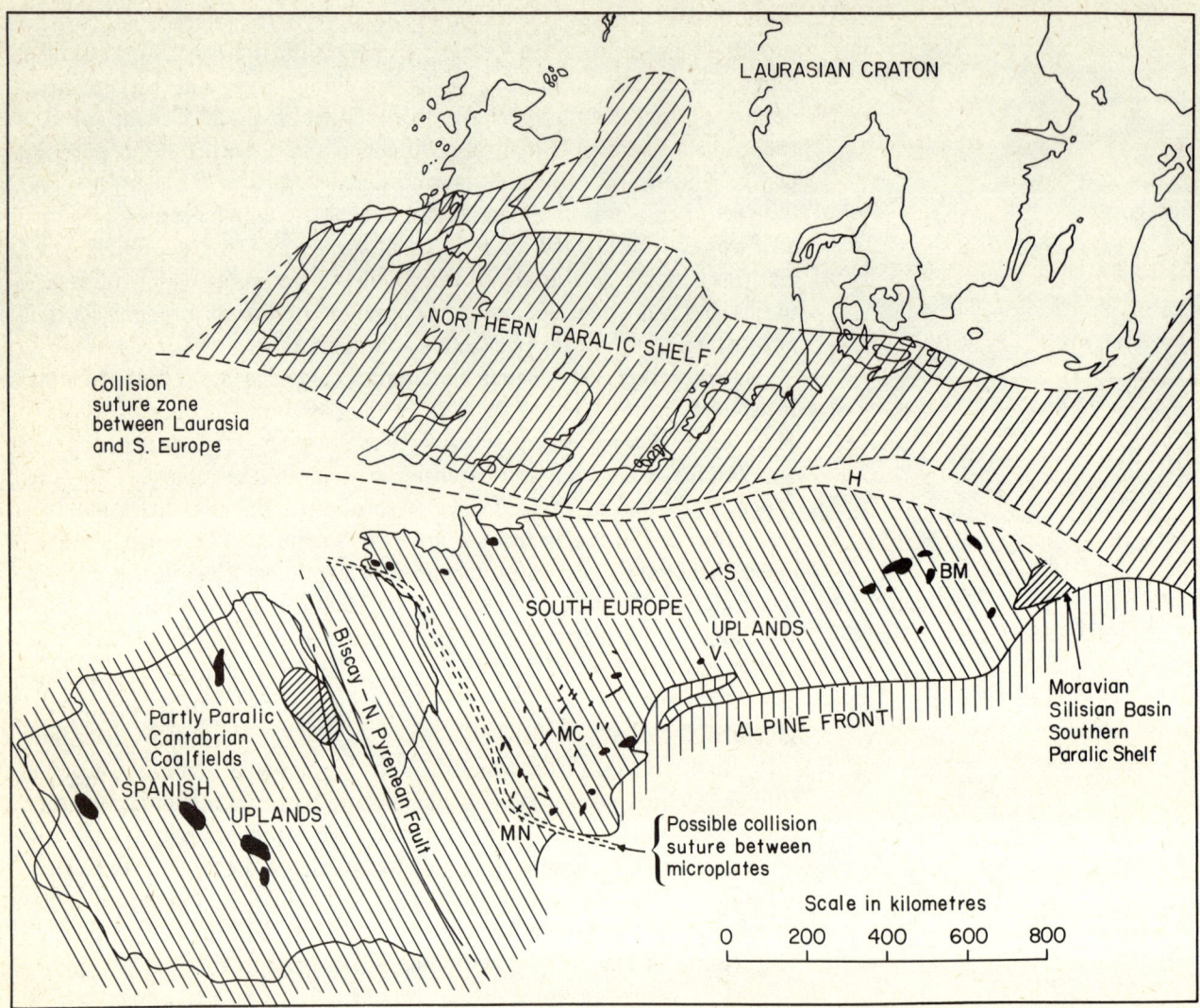

Fig. 5.2. Western Europe showing the paralic and limnic regions of coal deposition. Limnic coal basins of south Europe shown in black. BM, Bohemian Massif; H, Harz; MC, Massif Central; MN, Montagne Noire; S, Saar; V, Vosges. (Partly after Johnson, 1978; and Arthaud & Matte, 1977.)

The formation of the Permo-Carboniferous continental basins of Bohemia and those of France was controlled by post-orogenic uplift and erosion of the European Hercynian ranges. Formation of an active Andean continental margin has already been suggested to explain this orogenic episode, but it may be more complicated than this. A major lineament running through south Brittany and possibly continuing between the Massif Central and the Montagne Noire (Fig. 5.2) has been described by Lafort (1977 & 1979). This zone is characterised by features that suggest a collision suture and it could be that the Upper Carboniferous-Permian region of continental deposition in France and Bohemia was partly the result of Dinantian collision between two microplates. More stratigraphical and structural evidence is needed to substantiate this suggestion.

Transcurrent motion between the south Europe and African plates towards the end of the Carboniferous is believed to have formed a right lateral shear zone in south Europe and north Africa according to Arthaud & Matte (1977). This deformation consisted of wrench-faulting in response to horizontal compression. Sedimentary basins and troughs can form along major transcurrent faults. They are related to bends and irregularities in the fault-zone traces and occur at their terminations. Tertiary pull-apart basins and depressions along the line of the San Andreas fault, California, contain sequences of marine and terrestrial sediments up to 12 000 m thick (Crowell, 1974). Similar Late Carboniferous age basins are developed along wrench faults in southern France. The Stephanian Ales limnic coal basin lies on a left-lateral wrench fault which dies out southwards in a series of splay faults; sedimentary basins are developed in the splay-faulted area (Arthaud & Matte, 1977). Upper Carboniferous coal basins, some partly paralic, are also associated with the Biscay-North Pyrenean wrench fault in the Cantabrian Mountains, northern Spain (Heward & Reading, 1980). The nature of the deformation and uplift, with lack of volcanicity, suggests dominantly transgressive strike-slip movement during basin formation. Early deposition is marine in these basins and it becomes progressively more continental in type as the basins develop further towards the south-west. Though not an important environment for coal formation, significant coal-bearing successions have been formed in basins associated with major transcurrent faults.

On the eastern seaboard of the USA there are Upper Carboniferous limnic coal basins very similar to those of Europe. The Fundy Basin of Nova Scotia (Fig. 5.3) contains fluviatile and fluvio-lacustrine coal basins of Upper Carboniferous age that developed in a post-orogenic setting after Devonian (Acadian) collision orogeny (Hacquebard, 1972; Keppie, 1977). In Massachusetts the Narragansett Basin (Fig. 5.3) is a structural basin of some 2460 km^2 where early to late Pennsylvanian (Upper Carboniferous) rocks unconformably overlie Cambrian or Precambrian basement. Deposits consist of sandstone, conglomerate, shale and coal seams and indicate a fluviatile environment of deposition. Late Carboniferous earth movements have metamorphosed the succession to lower greenschist facies. Other basins in New England are similar and they are thought to have been formed in a region of high relief following the Acadian (Devonian) collision orogeny (Skehan *et al.*, 1979).

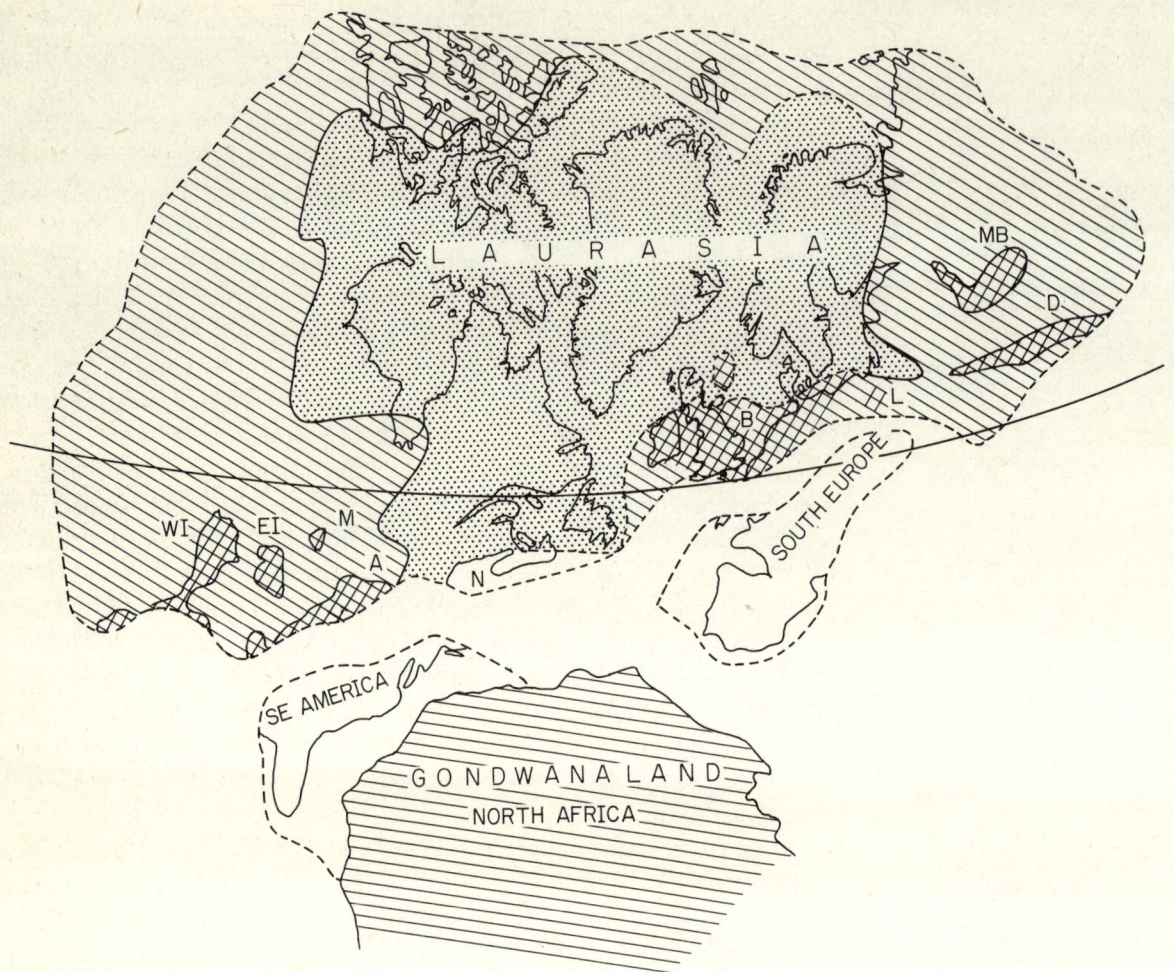

Fig. 5.3. Palaeogeographical interpretation of the Laurasian continent in the Middle Carboniferous showing the deduced position of converging continental masses and the location of intervening marginal basins. The south Nova Scotia plate (N) is fused to Laurasia. Carboniferous equator marked to the south of the British Isles. Major coal basins of Carboniferous age shown by double oblique shading: A, Appalachian; B, British Isles and North Sea; D, Donetz; EI, Eastern Interior Basin; L, Lublin Basin; M, Michigan Basin; MB, Moscow Basin; WI, Western Interior Basin. (Partly after Johnson, 1980.)

Similarity of the New England limnic coal basins with those of central Europe is striking and extends to their history of formation. Both regions developed as independent microplates lying on the seaward side of marginal basins south of the Laurasian continent. The New England-south Nova Scotia marginal basin closed forming Acadian orogeny during the Devonian and post-orogenic continental basins developed within the new mountain ranges. The central Europe marginal basin (Mid European Sea) closed with orogeny during the latter part of the Carboniferous period and post-orogenic lake basins formed from Westphalian through Permian within newly uplifted mountains (Figs 5.2 and 5.3). The development of coal seams in the Upper Carboniferous and Permian continental basins was largely dependent on favourable geographical position within the tropics and climate with adequate rainfall.

Coal

A major Tertiary continental yoked basin with vast coal deposits lies in North America in the east of the Rocky Mountains. The Rockies were formed during the late Mesozoic orogenesis and post-orogenic interior basins began to form in the Palaeocene. These basins were extensive and spread through the Dakotas, Wyoming, Montana, Saskatchewan and Alberta. Deposition was fluvial and includes friable yellow sandstone, grey shales and enormous deposits of low-grade coal (mainly sub-bituminous coal and lignite). According to Barrabe & Feys (1965) these American and Canadian prairie coals constitute 52% of total world reserves. Though this figure doubtless requires revision it does emphasise the size and importance of these deposits.

Gondwanaland coalfields

Outside Laurasia Late-Carboniferous, Permian and Triassic coalfields were developed in Angaraland and Gondwanaland in the northern and southern hemispheres, respectively. In both continents wide regions of subsidence developed during the Late Palaeozoic and Mesozoic and extensive coal measures were laid down. Both continents differ from Laurasia in straddling climatic zones outside the tropics through temperate and cool humid climates. Different plant floras would be expected in these temperate swamp forests and the conditions of coal formation would vary. Thus the Angara flora is distinct with index plants that are unknown outside Siberia in early Carboniferous times, but which spread rapidly in Asia towards the end of the period (Chaloner & Lacey, 1973). The Gondwanaland flora is equally distinct and completely restricted to the southern hemisphere. This flora developed with the evolution of a cold-tolerant class of plant, the Glossopteridae, which was able to colonise at the margins of ice cover during the Late-Palaeozoic continental glaciation of the southern hemisphere (Plumstead, 1973 a and b). The class is large and varied, from small shrubs to large trees, and all are deciduous. The group was highly successful and rapidly became dominant throughout Gondwanaland.

Comparison of the Gondwana coalfields of Africa with those of Laurasia has been made by Plumstead (1961) and Mackowsky (1968, 1975). The oldest coal deposits in Africa (Karroo group) are closely associated with the final waning phase of a vast Late-Palaeozoic glaciation or ice age; this is true of all the southern hemisphere Gondwanaland coalfields. The region of the Transvaal coalfields, one of the largest producers in the southern hemisphere, can be shown to be near the centre of ice-dispersal and has been suggested to be of relatively high altitude in Late-Palaeozoic times. It was far distant from the sea with the nearest marine deposits in South West Africa. The region formed a vast irregular continental interior basin at the time of coal formation in the Permian. The bottom coal seam is often interrupted by hillocks and ridges of pre-Karroo rocks around which coal-bearing sediments lie horizontally. Under these conditions deposition was very variable, gradients were high, coarse grits and feldspathic sandstones predominate. Sandstone partings and channel sandstones occur within coal seams. Linked with upland environment and adjacent valley glaciers the climate was cool and seasonal. Coal-forming plants show marked seasonal growth with a dormant period and tree trunks have well-defined annual rings. Thick autumnal deposits of single *Glossopteris* leaves also testify to the deciduous nature of the flora.

All these features produce particular characteristics in the coal seams (Plumstead, 1961; Mackowsky, 1975). Spores, pollen and leaf cuticles are concentrated in narrow bands owing to seasonal growth. Coarse mineral ash was derived from high-gradient mountain streams. Fine mineral ash, ubiquitous in southern hemisphere coals, is thought to be wind blown onto the leafless swamps during the winter months. Large fluctuations in the water table appear to be present with long spells of dry weather and much fusinite (natural charcoal) is present in the coal suggesting high oxidation of the peat during dry periods.

In the Transvaal coal basin there has been no lateral compression of the sequence and thus no true anthracites have been formed. The sediments are much as they were deposited, horizontal or with only slight dip. Late in the formation of the Karroo group a major phase of intrusions of dykes and sills had much effect on the coal seams. Heat alteration of the coals is widespread, sometimes to natural coke or even graphite. South African anthracite is heat-altered coal from the contact metamorphic zone of the dolerite intrusions. Intrusion of dolerite sills on the scale of the Karroo is almost unequalled in the world and it seems to be connected with regional sinking over a mantle hot spot. There is also a possible connection with early continental rifting between Africa and South America. Early stages of break-up involve subsidence to form structural depressions that become filled with continental deposits (Windley, 1977).

The lateral equivalent of the Karroo sequence continues in adjacent South America where very similar stratigraphy and faunas are developed and coal deposits are also found. In southern Brazil Late-Palaeozoic Gondwana coal occurs in the Paraná basin in deposits laid down in a continental interior depression formed after the retreat of a Late-Palaeozoic glaciation (Machado, 1975). The coal is of inferior quality, perhaps due to the cold-climate *Glossopteris* flora and to the slow sinking of the sedimentary basin. Both limnic and paralic coal basins of Gondwana type occur in Australia at the end of a Late-Palaeozoic glacial episode (Veevers & Evans, 1975). In India Gondwana coalfields are dominantly limnic in type, but there are important intercalations of marine strata. They are thus deemed paralic and are mentioned again in a later section.

THE PARALIC ENVIRONMENT

Continuous peat-formation and coalification on a regional scale can take place in the paralic environment on the edge of shallow shelf seas at Atlantic-type, aseismic continental margins. Large quantities of clastic sediment from an eroding hinterland are needed to build out a wide alluvial coastal plain and an extensive shallow continental shelf aids this development. Regional sinking of the shelf is necessary to allow a sedimentary succession be be built up and coastal swamp peats to be preserved. In the paralic environment marine deposits occur interbedded with deltaic, lacustrine and terrestrial sediments. A single marine horizon is sufficient to prove marine transgression over the shelf and the near proximity of the ocean. The paralic sediments are generally more travelled than those of limnic basins. Conglomerates composed of small pebbles and great thicknesses of sand, silt and clay build out wide deltas at or near to sea-level. These sediments prograde seawards from the shoreline. Repeated progradation of deltas leads to the development of cyclic sedimentation with the cyclothem coal, shale, silt and sand repeated many times in long sequences. The typical Carboniferous Coal Measures cyclothems of Britain and similar Pennsylvanian cyclothems of the USA have the general sequence: 1 coal; 2 marine or lacustrine limestone and/or shale; 3 shales; 4 interbedded siltstone and fine-grained sandstone; 5 sandstone with seatearth at the top. Similar sequences are widespread elsewhere in the stratigraphical column where paralic coal-swamp conditions occur. The great lateral continuity of these paralic delta-swamp deposits is demonstrated by strata such as the Pittsburg coal, Pennsylvanian (Upper Carboniferous) of the Appalachian Basin, USA, which is found over a region of 20 720 km² (8000 square miles). Coal Measures marine bands such as the Aegiranum and Vanderbeckei of Britain can be correlated reliably with the Aegir and Katharina Marine Bands of Germany over a distance of 800 km. In fact correlation of Upper Carboniferous paralic delta-swamp deposits along the entire southern margin of the Laurasian continent from the USA eastwards to the Donetz basin in Russia can be made with some precision (Ramsbottom et al., 1978).

Carboniferous paralic coal basins are particularly well developed in the British Isles, North Sea and across northern Europe to Poland.

Chapter 5

This region was a broad, shallow marine shelf on the Precambrian and Palaeozoic Baltic Shield. The shelf stretched from a coastline adjacent to the Scottish Highlands, the ancient Caledonian Mountain chains across Britain to Devon (south-west England) where it ends against a marine trench, the Culm trough. The shelf was structurally unstable and tended to sink differentially with negative grabens and half-grabens separated by positive horst blocks. Tensional stress in the upper crust, probably varying and impersistent in amount, affected the shelf region throughout the Carboniferous and is believed to have had much influence on downwarping and basin formation (Johnson, in press). Coal formation started in Dinantian (Lower Carboniferous) times and as the succession is traced seawards across the shelf it enters progressively higher in the sequence. Towards the middle of the Upper Carboniferous coal formation ended in the north and conditions changed to an arid redbed environment, the Barren Red Measures. Again the entrance of the arid interior coastal plain conditions rises seawards in the succession. In southern England and South Wales coal-swamp conditions persist, with interbedded red beds, to the top of the Carboniferous; in this area high in the Westphalian. This lateral variation in facies suggests progradation of deltaic sediments southwards from the Scottish uplands across the shelf. But the simple sedimentological picture is complicated by positive structural blocks and sinking basins. Low relief across the shelf throughout the period of Carboniferous deposition is suggested by widespread and persistent limestones and marine bands that extend northwards into Scotland. The presence of these marine bands distinguish the coal-bearing sequence as paralic rather than limnic.

Tensional stress in the crust and its influence on basin formation has already been mentioned. This stress field can be linked to contemporary plate movement because the subduction zone dipping south beneath South Europe is believed to have been continuous with the British shelf. Tensional stress was set up by the pull of the downgoing slab being translated to the continental crust trailing behind it (Johnson, in press). A structural pattern of this type has been recognised elsewhere such as in the Pre-Devonian of the northern part of the Iberian Peninsula (Riberto, 1974) and the Tertiary of New Caledonia (Paris & Lille, 1977).

The broad environment of Carboniferous paralic coal-swamp deposition in Britain suggests that it lies on the northern passive or aseismic margin of a marginal basin on the south side of the Laurasian continent (Fig. 5.3). This margin of the basin stretched from Ireland to Poland and includes the important British, North Sea, Belgian and West German coalfields. The basin closed towards the end of the Carboniferous with mild collision orogeny causing strong folding and thrusting of the offshore sediments (Johnson, 1978). Compression from the south also affected the South Wales and Radstock coalfields on the edge of the shelf at this time and influenced the formation of high-rank coals and anthracite in South Wales. Preservation of the greater part of the paralic shelf intact, with little or no tectonism of the cover of Carboniferous rocks, confirms the closure of a relatively narrow seaway of the marginal basin rather than the closing of a wide ocean (Le Pichon & Blanchet, 1978).

In south-west Laurasia, now North America (Fig. 5.3), similar conditions of deposition and structural history to those of north-west Europe are found. Paralic coal measures enter in the Pennsylvania (Upper Carboniferous) and extend through into the Permian. From a shoreline in the north, on the margins of the Canadian Shield, clastic sediment was transported across a broad unstable shelf that was intermittently invaded by transgressive seas from the south and east. Cycles of sedimentation, similar to those of Europe, were laid down. The shelf was divided into sinking basins and positive structural regions or blocks which were active during the time of deposition. Four major coalfields, the Mid-Continent, Illinois, Michigan and Appalachian, in eastern United States are the result of this structural framework. The Appalachian coal basin terminates eastwards against the Cincinnati Arch structural axis, and westwards is involved in the Appalachian fold mountains. The Pennsylvanian anthracite field in eastern Pennsylvania is part of the Appalachian coal basin that was folded during the Alleghanian Permian orogeny, the final phase of formation of the south Appalachian chains. This folding and the resulting pressure converted bituminous coal into anthracite. According to Dunbar & Waage (1969) the Anthracite coalfield has produced almost one quarter of the total output of coal in North America. Coal seams are thick with the Mammoth averaging 10.5-12 m and reaching 34 m at one location owing to overfolding. The time of folding can be dated as Permian because the basal Permian Dunkard Group is folded with the Pennsylvanian and is unconformably overlain by Trais.

The Appalachian Basin appears to have a structural setting rather similar to western Europe, and to be a paralic basin on the south side of Laurasia (Fig. 5.3). Stratigraphical and structural interpretation of the southern Appalachians requires more study before a detailed sequence of events can be reconstructed, as in Europe, but the basin seems to have closed to a narrow sedimentary trough by Carboniferous times and finally closed with the Alleghanian folding in the Permian. The Alleghanian episode is correlated with the collision of Africa Gondwanaland with North America (Dewey & Burke, 1973).

Chapter 5

Paralic coal basins were also developed on the Russian Platform in eastern Laurasia during the Carboniferous. The Moscow basin is a broad downwarp containing productive coal measures interbedded with thick Lower and Upper Carboniferous marine carbonates. To the east similar successions are developed in the Urals, fold mountains of Permian age that were formed by collision orogeny of the Angaraland (Siberia plate) with eastern Laurasia (Hamilton, 1970). In the south of the Russian Platform a large elongate downwarp developed on the north and east side of the Ukraine Shield in Devonian times. This basin, called the Donets aulacogen, formed initially as a graben on Precambrian structural lines. It contains a thickness of some 10 000 m of Carboniferous sediments composed of interbedded terrestrial and marine deposits with well-developed coal seams. Cyclic sedimentation reminiscent of the cyclothems of Europe and North America continue in the succession of the Donetz basin to the base of the Permian. Subsidence was of epeirogenic type, but could possibly be linked to the closely adjacent and eroding Ukraine Shield in which case the Donetz would be partly a yoked basin.

Outside Laurasia paralic coal basins developed widely in the Gondwanaland continent during Carboniferous and Permian times. In peninsula India marine transgressions have been recorded in Permian age coal measures which are predominantly fresh water in environment (Bharadwaj, 1969); the exact palaeogeographical setting here is uncertain. Interbedded marine and terrestrial deposits with coal seams in West Africa occupy a depression that may be linked with crustal stress developed during the preliminary phase of the opening of the South Atlantic in the same way as the Karroo basin (p. 110). The Sydney Basin of New South Wales, Australia, formed to the south of a major landmass and extensive deltaic sedimentation, produced coal measures with individual coal seams that can be traced laterally for over 160 km without reaching the limits (Brown *et al.*, 1968). Other examples of paralic coal basins can be described from the post-Palaeozoic part of the stratigraphical column, and it is noteworthy that peat deposits are being formed today almost at sea-level in various parts of the world. The Everglades and Dismal swamps of southeastern USA are examples of actively forming paralic swamps. And in the Niger and Mississippi deltas both swamps and buried peat deposits give conditions which could eventually provide sedimentary rocks with coal seams.

5.4 Coal formation and plate tectonics

A hot, humid climate in the low latitudes allowing rapid and continuous growth of luxuriant vegetation is one of the main requirements for the development of swamp peat on a regional scale. For coalification to take place gentle downwarping of the region and burial of the peat with younger strata is essential. With gentle subsidence coal seams and their associated sediments form a succession of almost horizontal strata. In the younger part of the stratigraphical column deposits may be found in this position, but it is much more usual for them to be disturbed by later earth movements. Originally almost horizontal strata are usually compressed and arched into folds and depressions and dislocated by faults and thrusts. With folding and faulting some strata gain potential for deeper burial within the earth's crust in concave synclinal folds and down-faulted blocks. Conversely, in convex or anticlinal folds and in uplifted blocks strata acquire potential for subsequent erosion. As a result, prolonged erosion of folded strata can lead to stripping of the uplifted regions and the preservation, possibly with deeper burial, of the protected areas. Separate coalfields are formed in this way in a region throughout which coal-formation took place originally. At the same time pressure or dynamic metamorphism owing to deep burial in synclinal basins, and due to folding, causes changes in coal seams towards the economically important high-rank coals and ultimately to anthracites.

In various ways plate movements, both rifting, approaching and after collision, provide mechanisms for basin formation and regional downwarping where, given the correct climate and environment, coal formation may take place. The early stages of continental rifting can form wide marginal basins in which both limnic and paralic coal basins can develop. Basins formed by aborted rifting, such as has been claimed for the North Sea Basin, can form important centres for the deposition of coal. But probably most important are the basins developed on the edge of continents with stable Atlantic-type margins where swamp development can spread on a regional scale. Even in this structural setting downwarping and basin formation can be influenced by tensional stress produced in the upper crust by active plate movement; this has been described with respect to the Carboniferous paralic coal basin of north-west Europe (p. 112). Post-orogenic intermontane interior basins are formed after collision orogeny and after orogeny of Andean type. The limnic coal basins that form under these conditions can be the repository of significant reserves of coal. Finally plate motions at active margins, and in collision, provide compressive forces to produce folding and the formation of coalfields and can initiate the pressure metamorphism required to form high-rank coals. Differing from hydrocarbons, coal is immobile, it can only be lost by uplift followed by erosion or by dynamo-thermal metamorphism in events, such as major continent to continent collision.

Chapter 5

References

Anderson, T.A. (1975) Carboniferous subduction complex in the Harz Mountains, Germany. *Bull. geol. Soc. Am.*, **86,** 77-82.

Arthaud, F. & Matte, P. (1977) Late Palaeozoic strike-slip faulting in southern Europe and northern Africa: Result of a right-lateral shear zone between the Appalachians and the Urals. *Bull. geol. Soc. Am.*, **88,** 1305-1320.

Barrabé, L. & Feys, R. (1965) *Géologie du charbon et des Bassins houillers.* Masson & Cie, Paris.

Bharadwaj, D.C. (1969) Lower Gondwana formations. *C.R. 6th Congr. Int. Strat. Geol. Carbonif., Sheffield,* 1967, 1, 255-278.

Britten, R.A., Smyth, M., Bennett, A.J.R. & Shibaoka, M. (1975) Environmental interpretations of Gondwana Coal Measure Sequences in the Sydney Basin of New South Wales. In *Gondwana Geology* (Ed. K.S.W. Campbell) Third Gondwana Symposium, Canberra, 1973, pp. 233-247.

Brown, D.A., Campbell, K.S.W. & Cook, K.A.W. (1968) *The geological evolution of Australia and New Zealand.* Pergamon Press, Oxford.

Chaloner, W.G. & Lacey, W.S. (1973) The distribution of Late Palaeozoic floras. In *Organisms and continents through time* (Ed. N.F. Hughes) Spec. Pap. Palaeont., 12, pp. 271-289.

Creaney, S. (1980) Petrographic texture and vitrinite reflectance variation on the Alston Block, N.E. England. *Proc. Yorks. geol. Soc.*, **42,** 553-580.

Crowell, J.C. (1974) Sedimentation along the San Andreas fault, California. In *Modern & Ancient Geosynclinal Sedimentation* (Eds R.H. Dott & R.H. Shauer) Spec. Publ. Soc. econ. Palaeont. Miner., Tulsa, 19, pp. 292-303.

Dewey, J.F. & Burke, K.C.A. (1973) Tibetan, Variscan and Precambrian basement re-activation: products of continental collision. *J. Geol.*, **81,** 683-92.

Dunbar, C.O. & Waage, K.M. (1969) *Historical Geology* Third Edition. Wiley, New York.

Gignoux, M. (1955) *Stratigraphic Geology.* English translation of 4th French edition, 1950. Freeman, San Francisco.

Hacquebard, P.A. (1972) The Carboniferous of Eastern Canada. *C.R. 7th Congr. Int. Strat. Geol. Carbonif.*, Krefeld 1971, 1, pp. 69-90.

Hamilton, W. (1970) The Uralides and the motion of the Russian and Siberian Platform. *Bull. geol. Soc. Am.*, **81,** 2553-76.

Heward, A.P. & Reading, H.G. (1980) Deposits associated with a Hercynian to late-Hercynian continental strike-slip system, Cantabrian Mountains, Northern Spain. In *Sedimentation in oblique-slip mobile zones* (Eds H.G. Reading & P.F. Ballance) Spec. Pub. Int. Assoc. Sedimentologists, 4, pp. 105-125.

Holub, V.M. (1977) The Permo-Carboniferous continental basins of Bohemia and Moravia. In *Symposium on Carboniferous Stratigraphy* (Eds V.M. Holub & R.H. Wagner) Geol. Surv. Czechoslovakia, Prague.

Johnson, G.A.L. (1978) European plate movement during the Carboniferous. In *Evolution of the Earth's Crust* (Ed. D.H. Tarling) Academic Press, London, pp. 343-360.

Johnson, G.A.L. (1980) Carboniferous geography and terrestrial migration routes. In *The terrestrial environment and the origin of Land Vertebrates* (Ed. A.L. Panchen) Academic Press, London, pp. 39-54.

Johnson, G.A.L. (In press) Carboniferous sedimentary cycles in Britain controlled by plate movements. *C.R. 9th Congr. Int. Strat. Geol. Carbonif.*, Urbana, 1979.

Keppie, J.D. (1977) Tectonics of Southern Nova Scotia. *Nova Scotia Dept. of Mines,* paper 77-1, pp. 1-34.

Lefort, J.P. (1977) Possible 'Caledonian' subduction under the Domnonean domain, North Amorican area. *Geology*, **5,** 523-526.

Lefort, J.P. (1979) Iberian-Armorican arc and Hercynian orogeny in western Europe. *Geology*, **7,** 384-388.

Le Pichon, X. & Blanchet, R. (1978) Where are the passive margins of the western Tethys Ocean? *Geology*, **6**, 597-600.

Machado, E.R. (1975) Gondwana Coal in Southern Brazil. In *Gondwana Geology* (Ed. K.S.W. Campbell) Third Gondwana Symposium, Canberra, 1973, pp. 271-293.

Mackowsky, M.-Th. (1968) European Carboniferous Coalfields and Permian Gondwana Coalfields. In *Coal and coal-bearing strata* (Eds D. Murchison & T.S. Westoll) Oliver and Boyd, Edinburgh, pp. 325-345.

Mackowsky, M.-Th. (1975) Comparative petrography of the Gondwana and northern hemisphere coals related to their origin. In *Gondwana Geology* (Ed. K.W.S. Campbell) Third Gondwana Symposium, Canberra, 1973, pp. 195-220.

Paris, J.F. & Lille, R. (1977) New Caledonia: Evolution from Permian to Miocene. Mapping data and hypotheses about Geotectonics. In *International symposium on Geodynamics in south-west Pacific Noumea (New Caledonia).* Editions Technip., Paris, pp. 195-208.

Plumstead, E.P. (1961) The Permo-Carboniferous Coal Measures of the Transvaal, South Africa—an example of the contrasting stratigraphy in the Southern and Northern Hemispheres. *C.R. 4th Congr. Int. Strat. Geol. Carbonif.*, Heerlen, 2, pp. 545-550.

Plumstead, E.P. (1973a) The Late Palaeozoic *Glassopteris* Flora. In *Atlas of Palaeobiogeography* (Ed. A. Hallam) Elsevier, Amsterdam, pp. 187-205.

Plumstead, E.P. (1973b) The Enigmatic *Glassopteris* Flora and Uniformitarianism. In *Implications of Continental Drift to the Earth Sciences.* (Eds D.H. Tarling & S.K. Runcorn) Academic Press, London, 1, pp. 413-424.

Raistrick, A. & Marshall, C.E. (1939) *The nature and origin of coal and coal seams.* English Universities Press, London.

Ramsbottom, W.H.C., Calver, M.A., Eager, R.M.C., Hodson, F., Holliday, D.W., Stubblefield, C.J. & Wilson, R.B. (1978) *A correlation of Silesian Rocks in the British Isles.* Geol. Soc. Lond. Special Report No. 10, 82 pp.

Ribeiro, A. (1974) *Contribution à l'étude tectonique de Tras-os-Montes oriental.* Serviços Geológicos de Portugal, Memoria 24, 168 pp.

Skehan, J.W., Murray, D.P., Hepburn, J.C., Billings, M.P., Lyons, P.C. & Doyle, R.G. (1979) *The Mississippian and Pennsylvanian (Carboniferous) systems in the United States—Massachusetts, Rhode Island and Maine.* U.S. Geol. Surv. Prof. Pap. No.1110-A-L, A1-A30.

Smith, D.B. & Francis, E.A. (1967) *Geology of the country between Durham and west Hartlepool.* Mem. Geol. Surv. Gt. Brit., H.M.S.O.

Veevers, J.J. & Evans, P.R. (1975) Late Palaeozoic and Mesozoic History of Australia. In *Gondwana Geology* (Ed. K.S.W. Campbell) Third Gondwana Symposium, Canberra, 1973, pp. 579-607.

Windley, B.F. (1977) *The Evolving Continents.* Wiley, London.

Chapter 6

Ore Deposits of the Ocean Crust

J.R. CANN

6.1 Introduction

The oceanic crust is intimately related to plate tectonics, since it is constantly created and destroyed during plate tectonic movements. During the creation of oceanic crust at mid-ocean ridges, and its passage across the ocean basins towards its destruction in the oceanic trenches, several different kinds of metallic ore deposit are created. All of these kinds of ore deposit are visible in ophiolite complexes, the fragments of crust formed by ocean-floor spreading, and brought above sea-level by mountain building. Some of them have now been discovered in the ocean crust itself, where the clearest evidence is found of how these deposits form. The two most important kinds of ore deposit related to ocean-floor spreading are podiform chromite ores, mined in ophiolite complexes and the source of much of the world's chromite, and massive copper and zinc sulphide ores, mined in ophiolites and of potential economic interest on the ocean floor, too. The manganese nodules of the ocean floor, also of considerable interest though not yet exploited, will be consideed only in outline here, since they, or at any rate the more economically viable of them, form without much participation from plate tectonics. This chapter will first consider processes during the formation and evolution of the ocean crust, then the magmagenic ores, followed by the hydrothermal ores and finally manganese nodules.

6.2 Oceanic crustal processes

The oceanic crust is now known to form during the spreading apart of lithospheric plates, which are rigid in relation to the underlying asthenosphere. As the plates move apart, the soft asthenosphere wells up from below and is progressively converted to lithosphere to form part of the plates. During the rising of the asthenosphere it melts partially to produce basalt, which is eventually erupted onto the ocean floor, where newly erupted lava can be seen at the mid-ocean ridge crests. Fig. 6.1 shows the general process schematically.

The basalt magma probably forms by decompression melting of the rising asthenosphere, which is probably close to its melting point at depth. The melting point of basalt increases with depth several times

Chapter 6

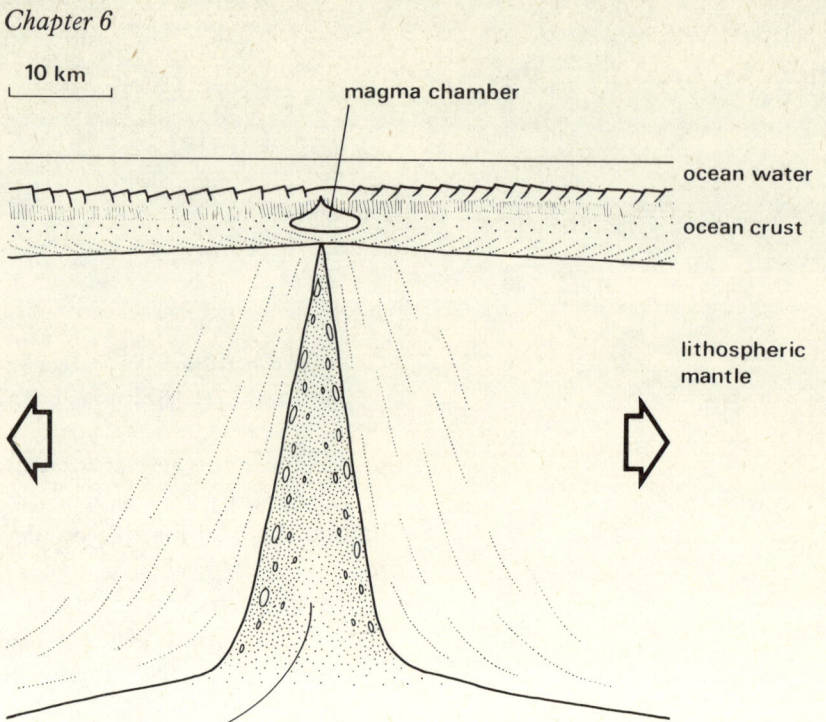

Fig. 6.1. Schematic cross-section of ocean-floor spreading. Plates of lithosphere move apart, and rising asthenosphere melts before transforming to lithosphere.

faster than the adiabatic gradient, so that asthenosphere close to, but below, the basalt melting point at depth will soon intercept it as it begins to rise. It seems, although this is conjectural, that the initial centimetre-sized drops of liquid aggregate during shearing and eventually migrate up through the overlying asthenosphere. The removal of liquid converts the asthenosphere to lithosphere, which adheres to the edge of the plate and spreads away with it. The residual mantle has a very characteristic composition and appearance. It is a gneissic peridotite, a harzburgite, composed essentially of 70% olivine, 25% enstatite and 5% chromite in the form of small clusters of crystals or vermiform patches.

Gneissic harzburgite of this kind is found in the ocean crust, generally highly serpentinised, in places where it seems to have intruded upwards as cold salt-dome-like diapirs along faults. It is also found to underlie the gabbroic parts of ophiolite complexes, where its field relations can be studied readily. The composition of the harzburgite shows that basaltic magma has been removed from it effectively, and its nature as a residue from partial melting is suggested by the constant proportions of the phases it contains.

In ophiolites the harzburgite can be seen to contain a characteristic set of veins and inclusions, principally composed of pyroxenite and dunite.

These bodies sometimes cross-cut the foliation in the harzburgite, but structural studies show that they have participated in at least some of the deformation of the host harzburgite. It is within the dunite pods and veins that the chromite deposits are found, and this part of the process will be considered further in the next section.

After passing through the mantle, the rising blobs of magma enter a crustal magma chamber, and it appears that the oceanic crust is generated by the interplay of processes within the magma chamber. On one hand, magma is extruded from the chamber upwards to form dykes and lavas on the ocean floor. On the other hand, crystals from the magma chamber accumulate on its floor to give layered gabbros. The magma chamber may also cool upwards to give a gabbroic carapace above it. This complex of processes is maintained by injection of new blobs of magma from below, and by continuous lateral spreading, which removes lavas, dykes and gabbros to the sides. Ideally it operates continuously, with the size of the magma chamber varying with spreading rate, but in practice, especially at slow spreading rates, fluctuations in the supply of magma apparently allow the chamber to dry up from time to time.

The resulting crustal structure, based upon observations in the oceans, necessarily rather indirect, and mapping of the ophiolite complexes, is shown in Fig. 6.2. Note how the magma chamber, steady in size, is able to generate a great quantity of ocean crust by the process of spreading.

It is now necessary to turn to the cooling of the ocean crust in order to understand the way in which hydrothermal processes form part of the picture. At the point where asthenosphere turns into lithosphere, the whole lithosphere has the temperature of molten basalt, approximately 1200°C. It is cooled in two ways, by conduction and, in the upper part, by convective penetration of sea water into it. The relative contribution of the two can be estimated by comparing the theoretical conductive cooling curve for such a slab of lithosphere with observations of actual conductive flow through the ocean floor. A considerable difference is observed, which can be ascribed to penetrative convection. Measurements of low heat-flow are probably areas where the convection is being recharged from the oceans, while the high measurements only barely reach the theoretical level. In support of this indirect evidence, hot springs have been observed on the ocean floor, both by deep-towed vehicle and also from submersibles. Most of these have been discharging at relatively low temperatures, around 10°C, but at one place near 21°N on the East Pacific Rise the US submersible *Alvin* found vents discharging at 350°C. The water of these vents was sooty-black with finely divided iron, zinc and copper sulphides precipitating as the hot spring water cooled in the sea.

Chapter 6

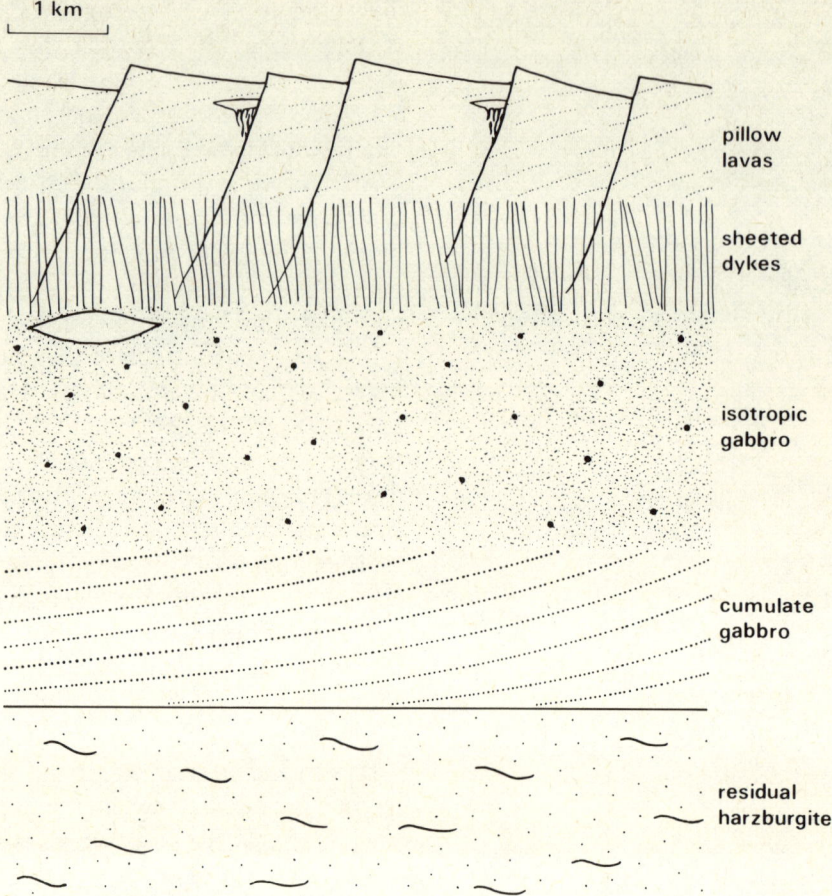

Fig. 6.2. Lithological cross-section of the ocean crust, based on results of dredging and drilling, and observations of ophiolite complexes.

Such high-temperature springs are apparently restricted to within a kilometre or so of the axis of spreading. The cooling is so efficient that after that initial phase of high-temperature cooling, the crust never becomes as hot again. Subsequent cooling seems to involve only low-temperature springs, although this still forms an important part of the process of cooling the oceanic lithosphere. The low-temperature springs are apparently not associated with sulphide deposits, but can be surrounded by mounds of iron and manganese oxides, and iron-rich clays, which have now been found in several places on the ocean floor. Unfortunately, none of these oxide deposits contain enough copper or other base metals for them to be economically attractive.

As the crust ages on its path across the ocean basins, low-temperature convective cooling remains an important process while outcrops of basaltic rocks remain, but once they are covered with sediment the rate of convection open to the sea slows down. Convection may continue in the relatively permeable basalt beneath the sea floor but it is essentially closed. The age at which this happens depends on the sedimentation rate and the roughness of the basement, both of which can vary by as much as two orders of magnitude. In some parts of the equatorial Pacific the crust can become sealed as soon as 5 million years after forming while the lithosphere is still hot, while in parts of the Atlantic and Indian Oceans sealing may not happen for 50 million years.

6.3 Magmagenic ore deposits

PODIFORM CHROMITE DEPOSITS

The most important class of magmagenic ore deposits associated with oceanic spreading centres is that of the podiform chromite deposits (Thayer, 1964). Very few pieces of chromite-rich rock have been discovered on the ocean floor, and it is only remotely likely that chromite will ever be mined from the ocean floor. However, podiform chromite deposits in ophiolite complexes are an important source of chromite; and an important exploration target. Chromite, a spinel with the ideal formula $FeCr_2O_4$, is an important raw material both for the manufacture of chromium metal and its alloys and also for the manufacture of refractory furnace linings. The particular usefulness of any deposit depends on the content of different minor elements which move the composition away from the ideal one.

A typical podiform chromite deposit is shown in Fig. 6.3. It consists of a central core, the pod itself, composed largely of chromite. Within the pod the chromite is generally coarse-grained, with a grain size in excess of 2 mm. Often the chromite occurs as the beautiful leopard ore, in which the chromite forms ovoid nodules up to several centimetres in diameter. In between the chromite grains or nodules, olivine, or serpentine formed from it, make up a minor part of the ore. The pods range in size from only about one ton up to several million tons. Of course, only the larger pods are economically viable, but it is important to emphasise the continuum of size.

The pods themselves are almost always enveloped in mantles of dunite or serpentinised dunite which are several times larger than the pods themselves. The dunite contains isolated large (2 mm) chromite grains, but little of any other mineral other than olivine or serpentine. The dunite envelope is a good local guide to the existence of chromite mineralisation, if the chromite itself is not exposed. Although the dunite appears structureless, this is because the olivine grains have little shape anisotropy. Petrofabric measurements show that the olivine does have a strong preferred orientation of its crystal structure, caused by rotation

Chapter 6

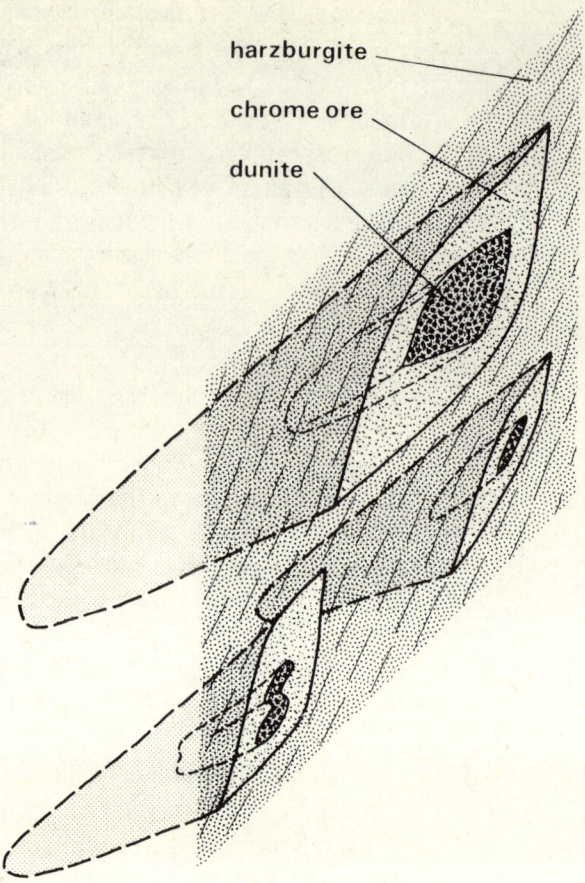

Fig. 6.3. A typical podiform chromium deposit seen as a block diagram.

of the grains during deformation towards orientations allowing more ready sliding of the ionic layers over one another. There has thus been an episode of deformation affecting the dunite (and also the chromite) after its formation.

Outside the mantled pod is the gneissic harzburgite forming the lower part of the ophiolite complex, representing sub-oceanic upper mantle, as described in the previous section. This is composed of olivine and enstatite, or their serpentinised equivalents, together with minor amounts of chromite. Within the harzburgite, chromite grains are much smaller than in the dunite, no larger than about 0.2 mm, and forming elongate clusters of grains parallel to the foliation in the harzburgite. The harzburgite has a characteristic orange-brown weathered surface covered with rough warty bumps which are the less rapidly weathering enstatites. This is in contrast to the dunites, which weather to a definitely lighter shade of yellow-buff (the dun colour after which dunite is named), and have a smooth weathered surface, showing scattered chromite grains. The fabric of the harzburgite is well marked by the elongation of the enstatite grains, which give it its characteristic gneissic appearance.

There seems to be a definite tendency for the larger sized chromite pods to occur near the top of the harzburgitic part of the ophiolitic suite, in the zone immediately beneath the gabbroic crustal rocks. This in turn provides a further guide to locating economic chromite deposits from further away. Exploration programmes should be concentrated within the zone of harzburgite adjacent to the gabbroic units provided the ophiolite has not been too much dissected by faulting. Within this zone, the occurrence of dunite provides a second exploration guide, which allows identification of limited zones of greater potential for detailed work.

There has been much debate in the past about how deposits of podiform chromite have arisen, much of it before any clear idea was available about how ophiolite complexes are formed, and the significance of their ultramafic members. These theories are reviewed at length in other books on ore deposits (see Park & McDiarmid, 1975; Stanton, 1972; Dickey, 1975). Appreciation of the geological environment in which such deposits have formed and increased knowledge of processes at spreading centres have led to a remarkable clarification of the picture, one of the most important contributions of plate tectonic ideas to ore geology. It is this clarified picture that is presented here (Fig. 6.4).

The harzburgite country rock has the appearance of residual mantle from which basalt has been extracted, as explained above. When the basalt had gone, the remaining liquid was in equilibrium with olivine enstatite and chromite, the solid phases in the harzburgite. However, the most basic basalts of the sea floor have phenocrysts of olivine and chromite, with no enstatite, as might be expected. The reason for this has been shown experimentally to be caused by change of pressure. If basalt magma, originally in equilibrium with harzburgite, rises into regions of lower pressure, it becomes oversaturated with olivine (the olivine phase volume expands), and the enstatite phase volume contracts (O'Hara, 1968). Thus pockets of rising magma precipitate olivine and chromite progressively as they rise.

Most of the olivine and chromite formed in this way seems to be left behind as thin streaks and small blobs, which are sheared out into veins during the deformation of the mantle beneath ridge axes. Occasionally, however, a rather larger than usual blob of magma may pause for a relatively long time, and thus generate a larger mass of chromite and the associated olivine as dunite surrounding it. The larger masses are also deformed, though not as intensively as the harzburgite, since it has already suffered deformation before the chromite-dunite pod was emplaced.

Chapter 6

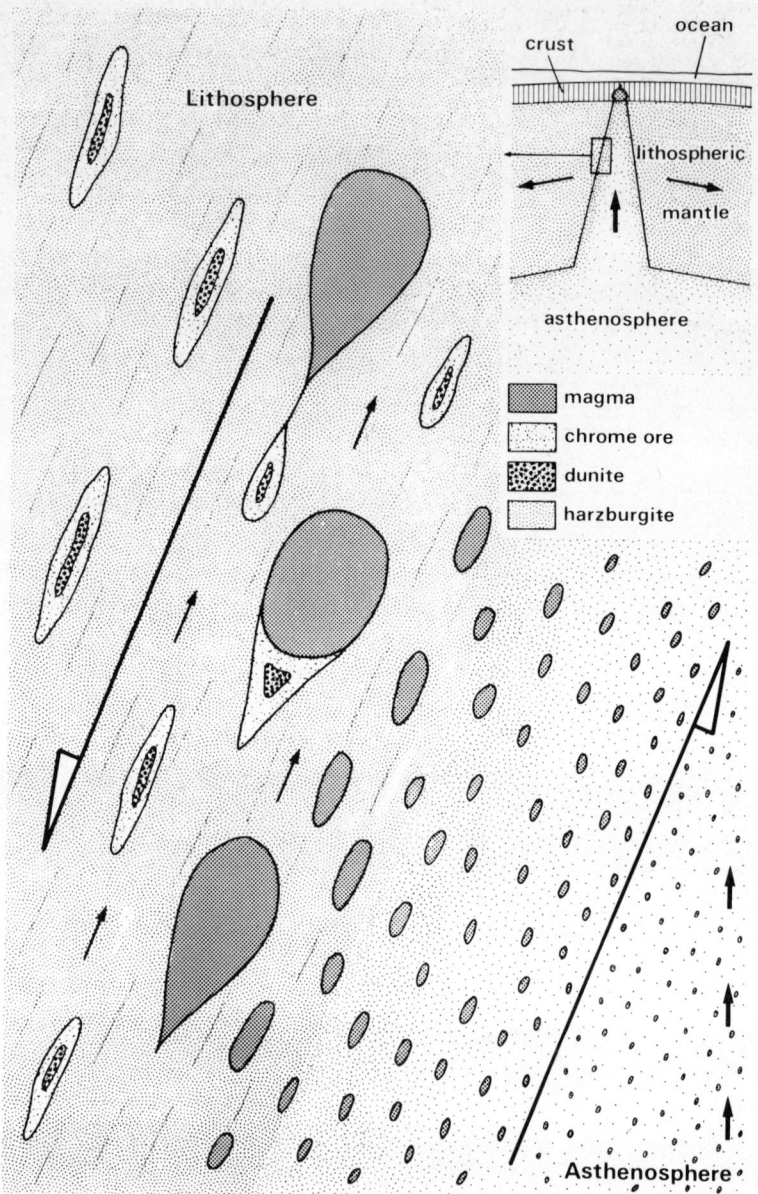

Fig. 6.4. Formation of podiform chromite during ascent of asthenosphere and segregation of basalt beneath mid-ocean ridges.

Larger chromite deposits may occasionally be formed in a rather different way. As the basalt magma rises upwards it eventually enters the crustal magma chamber from which the crust is generated. If the magma chamber has been in place for a long time, then the basalt magma in it is likely to have evolved rather far from its primitive composition, and to be precipitating plagioclase and augite, as well as olivine, and no chromite. However, especially on slow-spreading ridges, the magma chamber will occasionally freeze-up between widely spaced inputs. When that happens it will be refilled with primitive unevolved magma, which will at first precipitate olivine and chromite.

The first cumulates to form from a reborn magma chamber, lying immediately above the residual harzburgite, may thus contain economic deposits of chromite. Podiform chromite deposits thus seem to be indisputably magmatic in origin, and are closely associated with the processes by which basaltic magma is segregated from the mantle into magma chambers. The magmatic model accounts for the salient features of chromite deposits, and for their location in the upper part of the harzburgite units of ophiolite complexes. In addition it allows a more precise understanding of how to conduct a search for the chromite ore.

OTHER MAGMAGENIC ORES

Two other classes of ores have been recorded in association with ophiolite complexes, but neither so far in concentrations as economically attractive as in other settings. These are segregation deposits of platinum-group minerals and segregations of iron, copper and nickel sulphides.

6.4 Hydrothermal ores

OBSERVATIONS ON THE OCEAN FLOOR

The most striking occurrence of hydrothermal ores on the ocean floor is that in the Atlantis Deep of the Red Sea. Though high temperatures had previously been obtained in hydrocasts in the Red Sea, the deposits of the Atlantis Deep were discovered as a result of the observation of a horizontal boundary serveral tens of metres above the sea floor by a research ship's echo-sounder. Water samplers revealed a pool of brine, about ten times more saline than sea water, approximately saturated with sodium chloride, and at high temperature. This deep was eventually named the Discovery Deep after the research ship. Further study showed that it was kept filled by overflow from a neighbouring larger deep itself replenished by hot springs, the temperature of which is as high as 56°C. Coring in the floor of the Atlantis Deep revealed that its sediments were metalliferous, consisting of different assemblages of minerals, including manganese and iron oxides, iron and other sulphides and iron-rich clays, especially nontronite, and carbonates. Apart from iron and manganese, the sediments were enriched in other base metals, especially zinc and copper. The best estimates of the size of the deposit (Hackett & Bischoff, 1973) is that the basin contains about 200×10^6 tons of ore, containing about 3.2×10^6 tons of zinc and 0.8×10^6 tons of copper, mostly in the form of sulphides. The deposits of the basin and the brines have been described in an important book edited by Degens and Ross (1969). Although more complex theories were advanced in that book, principally on the basis of stable isotope information, it seems now to be generally agreed that the deposits form as a result of sea water becoming hypersaline by dissolution of the evaporites that form great thicknesses on the flanks of the Red Sea basin, the percolation of this brine into hot, newly formed basaltic oceanic crust in the central deep of the Red Sea and the dissolving of base metals from the basalts by the heated brine. When the brines emerge in the Atlantis Deep they precipitate the metals as they cool and mix with the overlying sea water. Distinct horizons of ore precipitation can be seen in the brine of these pools.

There is no doubt that the Atlantis Deep deposit qualifies as an ore deposit, both in terms of containing metals and also as an economic proposition. At present (1980), preparations are actively underway for its exploitation as a shared project between Saudi Arabia and the Sudan. The interest aroused by this deposit has led to intensive investigation of the Red Sea for further deposits. This search has been successful in finding other brine pools or places where they previously existed, but none has yet been shown to have the economic interest of the Atlantis Deep (Bignell, 1975). Certainly the process seems to have been reasonably common in the Red Sea. However, no similar brine pools have been found in other parts of the ocean, casting doubt on whether this is a universal phenomenon. The answer seems to be that it is universal, but that it is normally not accompanied by hypersaline brine formation, and, thus, that the brine pools are not present except in places where evaporites are abundant.

Within the open ocean basins, the first indications of hydrothermal activity came from studies of heat conducted through the sediment of the ocean floor. The recognition of the formation of oceanic crust by ocean-floor spreading allowed well-controlled calculation of expected heat-flow in crust of known age. Comparison with observation showed considerable discrepancy, and Lister (1972) pointed out that this could be well understood if a substantial part of the heat of newly formed lithosphere is convected out via hot springs through outcrops of rock, so that conductive heat-flow did not measure total heat-flow. As the sediment becomes thicker and the total heat-flow less, convective heat-flow becomes less and less important, until eventually almost all of the heat is removed by conduction through the sediments. However, hot springs should be particularly important in transferring heat in very young crust, since there the conductive heat-flow measured is frequently less than one tenth of the calculated flow.

Shortly after this realisation, hydrothermal deposits began to be recognised on the ocean floor. Examples of what appears to be the most common type have been discovered in several parts of the oceans now: in the crust generated at the Galapagos Spreading Centre in the Pacific; at two places on the Mid-Atlantic Ridge in the North Atlantic; and in the Gulf of Aden. The best-developed instance is that near the Galapagos Spreading Centre, about 300 km east of the Galapagos Islands. Here chains of mounds, 20-30 m high and of a similar width, were found strung out along some of the ridge-crest-parallel faults on crust about $0.1-0.2 \times 10^6$ years old (Corliss *et al.*, 1978). The mounds have a carapace of manganese and iron oxides and hydroxides covering layers of green, crumbly nontronitic clay. Warm water was observed flowing out through the mounds, although its warmth (about 10°C) was only relative to ocean bottom waters (about 2°C).

The other deposits show the same association of manganese and iron hydroxides with the ferric iron-rich clay, nontronite, and both minerals seem to have precipitated out of the warm spring water. However, all of these kinds of deposit are disappointingly poor in base metals. Levels of Cu, Ni, Zn and Co are, at most, a few hundreds of parts per million, often well short of this and considerably less than any economic levels. Their best analogy in ophiolites seems to be the umber deposits found at the top of the lava pile, which are exploited primarily as pigments (Robertson, 1976).

Recently submersible exploration of the crest of the East Pacific Rise, near 21°N, south-west of the mouth of the Gulf of California has discovered much more interesting hydrothermal activity. Here, very close to the spreading axis, circular vents, up to 1 m across, are seen discharging very hot hydrothermal solutions. The temperature of the solutions is about 350°C, and they emerge charged with black, finely divided sulphides, precipitated as the fluids cool and begin to mix with the deep ocean water, giving the vents the appearance of densely smoking factory chimneys. Around the vents, cylinders of sulphides are precipitated, composed of chalcopyrite, pyrrhotite, pyrite and sphalerite. Such springs, under the right conditions, could clearly give rise to sea-floor sulphide deposits of some size, though since they form so close to the ridge axis the deposits would very likely be covered by later lava flows during continued crustal accretion.

Similar high original temperatures were deduced for warm springs discovered near the axis of the Galapagos Spreading Centre, near the equator, from the geochemical variation of the emerging waters. However, these waters must have been diluted with cold sea water within the ocean crust, and any sulphides they contained must have been precipitated there.

Chapter 6

Clearly ocean-floor sulphides are only likely to be viable exploration targets if they lie at the sea floor itself. It is never likely to be economic to exploit sulphides covered by 100 m of basalt as well as 4 km of water, even though thousands of such deposits probably exist scattered within the upper part of the ocean crust.

SULPHIDES IN OPHIOLITES

Similar sulphide deposits have been exploited in ophiolites since prehistoric times, and still form an important prospecting target. The classic deposits of this type are those of the Troodos ophiolite in Cyprus, re-examined in detail in the 1960s by a United Nations team and, thus, now extremely well understood (Constantinou & Govett, 1972, 1973; Searle, 1972). The deposits average about 3 million tons, and 13 deposits ranging between 0.5 and 15 million tons of ore have been discovered to date in the volcanic unit of the Troodos complex. All but one of the deposits lie beneath a unit of upper pillow lavas, and above a further unit of pillow lavas. A schematic diagram of a typical deposit is shown in Fig. 6.5. The ore itself is mostly present as massive sulphide and is overlain by a unit of

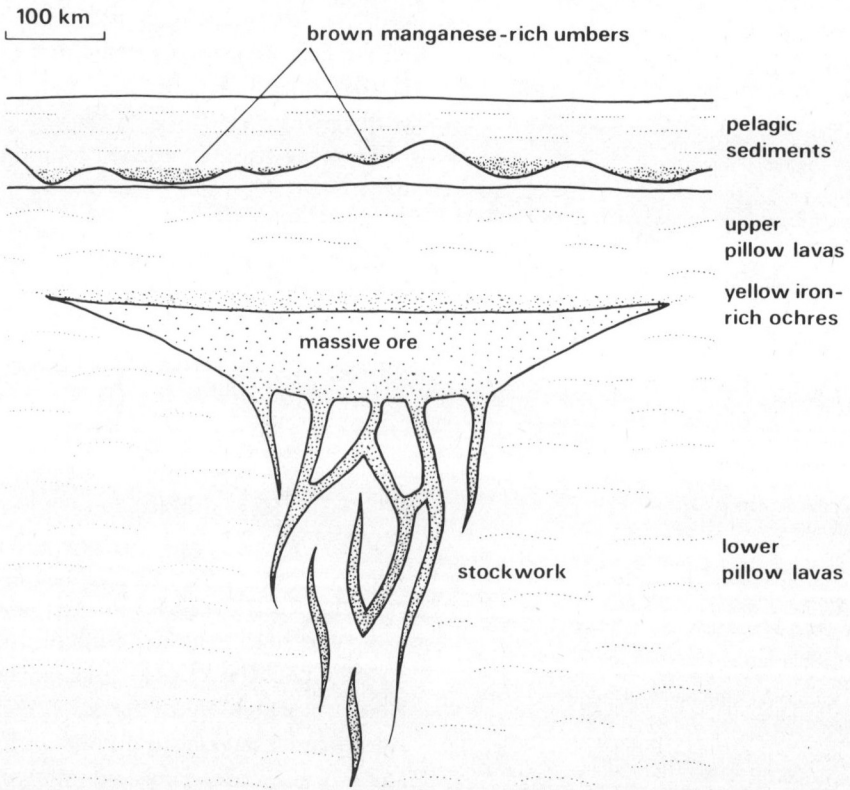

Fig. 6.5. Cross-section of Cyprus-type massive sulphide deposit, showing relations to surrounding rocks.

nontronitic clay and by material rich in iron oxides (the ochres) derived apparently from oxidation of the sulphides. The sulphide present is mostly pyrite, containing minor amounts of chalcopyrite and sphalerite, and other sulphides in smaller amounts. The ore was at one time exploited partly for its sulphur content, but interest now is principally in the copper, which ranges from a few tenths of a percent up to, occasionally, about 3%. Associated with the ore sometimes is the black 'devil's mud' which may contain significant gold values.

The saucer-shaped massive ore body is underlain by a stockwork of sulphide veins in brecciated pillow lava, itself metamorphosed to greenschist facies, implying temperatures in excess of 300°C. Within the stockwork quartz veins occur, and Spooner & Bray (1977) have shown that fluid inclusions within these veins imply temperatures of formation of about 350°C, from water of salinity close to that of sea water. In contrast, the upper part of the massive sulphides commonly show colloform banding, usually interpreted to imply a low temperature of deposition, and the overlying lavas show only low-temperature alteration.

All of this evidence suggests a close parallel with the high-temperature deposits from the ocean floor described in the previous section (Fig. 6.6). High-temperature fluids, of the same temperature as those seen in the active hot springs, have clearly risen through the underlying lava, debouching on what was then the sea floor, and depositing ores over a wide range of temperature as they are diluted and cooled. The covering of pillow lavas shows that all of this activity happened close to the spreading

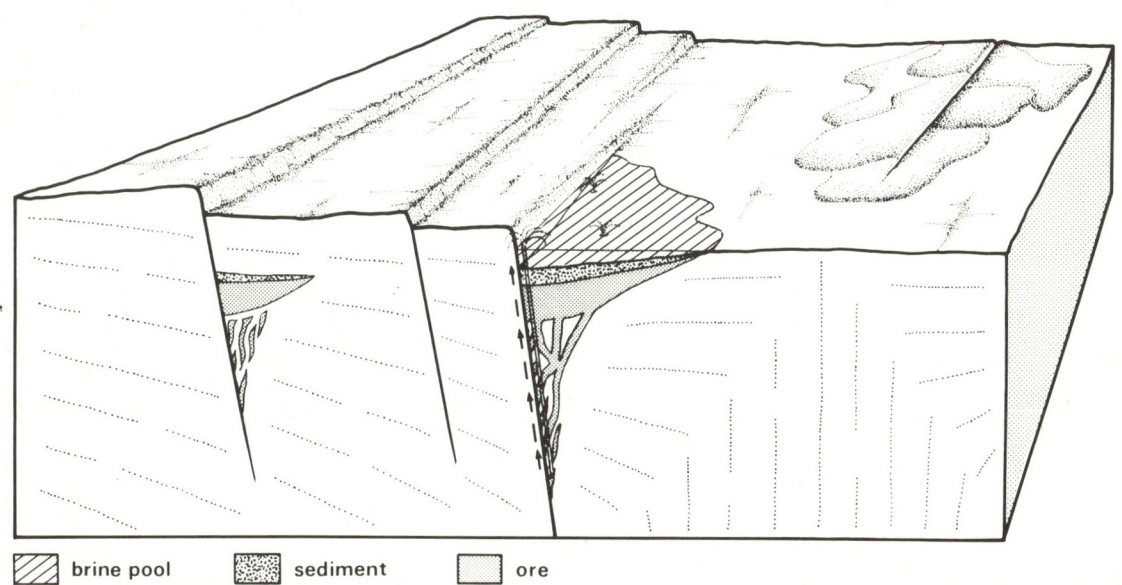

Fig. 6.6. Environment of formation of Cyprus-type massive sulphides at the mid-ocean ridge crest.

axis, no further away than could be reached by lava flows emitted there.

The frequency of occurrence of these deposits within the volcanic unit of the Troodos complex allows a computation of the frequency with which they may be met, uncovered, on the ocean floor (Cann, 1980). It appears that about one exposed deposit can be expected for every 100 km of ridge crest, and that the number of actively forming deposits would be substantially less than that. It seems a relatively fortunate circumstance that one such active deposit has been found so early in the exploration of the sea floor by submersible.

Similar sulphide deposits are not restricted in ophiolites to Cyprus. They have been mined in ophiolites in Newfoundland, Oman, the Philippines and are likely to be found in many more (Coleman, 1977).

6.5 Manganese nodules

Manganese nodules will only be described in outline here since, though they occur on the ocean floor, their occurrence is not closely related to plate tectonics. However, they are closely related in mineralogy to the low-temperature manganese oxide deposits of the mid-ocean ridges described above, and it is likely that a continuum of types of deposit exists between those clearly hydrothermal and the most enigmatic nodules of hydrothermal interest. Also related to these deposits are the more continuous layers of metalliferous sediment that occur at the base of the oceanic sediment column, and also, at times, within the sedimentary section.

The classic manganese nodule is roughly potato-sized and shaped, with a smooth convex outside surface, coloured dark brown. A section through a nodule shows it to have a nucleus of some kind—apparently any solid material will suffice, whether limestone, basalt or whale's earbone—surrounded by concentric finely laminated layers of manganese and iron oxides. Where abundant nodules may be scattered on the sea floor at densities of tens per square metre.

There is a complete transition from this classic kind of nodule, through larger nodules, reaching sizes of up to a metre across, which are characteristically asymmetrical in shape, to encrustations of manganese oxides on rock outcrops. Rates of accumulation can be measured from the rate of decay of ^{230}Th-activity with depth in the outer crust of the nodule, and fall into the range of a few millimetres per million years for nodules in deep ocean basins. Most nodules are thus accreting at rates three orders of magnitude less than sedimentation rates in the sediments on which they are found. They are apparently kept at the sea floor by bioturbation.

Not all nodules are of economic interest. Only those with high levels of Co, Cu and Ni, which may be in excess of 1%, can be considered seriously. Manganese alone does not make them worthwhile. Such nodules are restricted to limited areas of the ocean floor, of which the best known is on the floor of the Pacific south-east of Hawaii. Why there should be this sort of provincialism in nodule composition is not at all clear, though in general the more slowly accreted nodules are richest in trace metals, and the rapidly accreted manganese deposits of the ridge crests and the continental margins are typically very poor in base metals.

The origin of manganese nodules is still a subject of controversy. In particular it is not clear how much of the metal content of each nodule is derived from within the underlying sediment and basalt, and how much from the ocean water itself. Certainly the trace element signature of the nodules has some very hydrogenous characteristics: some of the more abundant trace elements, such as lead, titanium and molybdenum, are known either to be very depleted in the ocean crust or to be immobile there, and the rare-earth pattern is in general very similar to that of ocean water. In addition, there is considerable speculation about the role of bacteria in nodule formation.

Further reading

Coleman, R.G. (1977) *Ophiolites: Ancient Oceanic Lithosphere?* Springer-Verlag, Berlin, Heidelberg and New York.

References

Bignell, R.D. (1975) Timing, distribution and origin of submarine mineralisation in the Red Sea. *Trans. Inst. Min. Metall.,* **84,** B1-B6.
Cann, J.R. (1980) Availability of sulphide ores in the ocean crust. *J. Geol. Soc. Lond.,* **137,** 381-384.
Coleman, R.G. (1977) *Ophiolites: Ancient Oceanic Lithosphere?* Springer-Verlag, Berlin, Heidelberg and New York.
Constantinou, G. & Govett, G.J.S. (1972) Genesis of sulphide deposits, ochre and umber of Cyprus. *Trans, Inst. Mining. Metal.,* **81,** B34-B46.
Constantinou, G. & Govett, G.J.S. (1973) Geology, geochemistry and genesis of Cyprus sulfide deposits. *Econ. Geol.,* **68,** 843-858.
Corliss, J.B., Lyle, M., Dymond, J. & Crane, K. (1978) The chemistry of hydrothermal mounds near the Galapagos Rift. *Earth Planet. Sci. Lett.,* **40,** 12-24.
Degens, E.T. & Ross, D.A. (1969) *Hot brines and recent heavy metal deposits in the Red Sea.* Springer-Verlag, New York.
Dickey, J.S. (1975) A hypothesis of origin for podiform chromite deposits. *Geochim. Cosmoch. Acta,* **39,** 1061-1074.
Hackett, J.P. & Bischoff, J.L. (1973) New data on the stratigraphy, extent and geologic history of the Red Sea geothermal deposits. *Econ. Geol.,* **68,** 553-564.
Lister, C.R.B. (1972) On the thermal balance of a mid-ocean ridge. *Geophys. J. R. astr. Soc.,* **39,** 515-535.
O'Hara, M.J. (1968) Are ocean floor basalts primary magma? *Nature,* **220,** 683-686.

Chapter 6

Park, C.F. & MacDiarmid, R.A. (1975) *Ore Deposits* 3rd Edition. W.H. Freeman and Co., San Francisco, pp. 224-231.

Robertson, A.H.F. (1976) Origins of ochres and umbers: evidence from Skouriotissa, Troodos Massif, Cyprus. *Trans. Inst. Mining Met.*, **85**, B245-B251.

Searle, D.L. (1972) Mode of occurrence of the cupriferous pyrite deposits of Cyprus. *Trans. Inst. Mining Metal.*, **81**, B189-B197.

Spooner, E.T.C. & Bray, C.J. (1977) Hydrothermal fluids of sea water salinity in ophiolitic sulphide ore deposits in Cyprus. *Nature*, **266**, 808-812.

Stanton, R.L. (1972) *Ore petrology*. McGraw Hill, New York, pp. 307-324.

Thayer, T.P. (1964) Principal features and origin of podiform chromite deposits and some observations on the Guleman-Soridag district, Turkey. *Econ. Geol.*, **59**, 1497-1524.

Chapter 7

Ore Deposits associated with Subduction

R.D. BECKINSALE & A.H.G. MITCHELL

7.1 Introduction and historical background

By about 1967 it was recognised that the distribution of calc-alkaline volcanism in arcuate belts above Benioff zones is related to subduction of oceanic lithosphere. Economic geologists did not begin to discuss the distribution of mineral deposits associated with these igneous rocks in terms of plate boundaries until several years after this, mainly because it was clear that the plate tectonic hypothesis could not be expected to help identify the precise location of an ore body. Five years later, by about 1972, the distribution of relatively young (Mesozoic and Cenozoic) sulphide ore deposits above contemporaneous Benioff zones had been noted by several authors (Sillitoe, 1970; Snelgrove, 1971; Guild, 1971) and hypotheses were proposed to relate the source of the metals to subduction-related processes. Subsequently the ore deposits associated with much older calc-alkaline rocks in Phanerozoic or earlier orogens have been interpreted using uniformitarian principles as related to subduction of oceanic lithosphere. Plate tectonic interpretation of the mineralisation associated with continent-continent or continent-island arc collison belts did not start until about 1974 with the recognition of the possibility of generating magmas in such collision settings.

Mineralisation associated with magmatic rocks in four tectonic settings is discussed in this chapter:
1 Magmatic arcs
2 Back-arc magmatic belts
3 Outer-arc magmatic belts
4 Foreland fold-thrust belts

The first three of these tectonic settings lie on the overriding plate margin while the fourth is confined to continental forelands in collision belts.

7.2 Magmatism and mineralisation at convergent plate margins related to tectonic settings

The four plate tectonic settings discussed in this chapter are illustrated in Fig. 7.1, and brief descriptions of these four settings are as follows.

Chapter 7

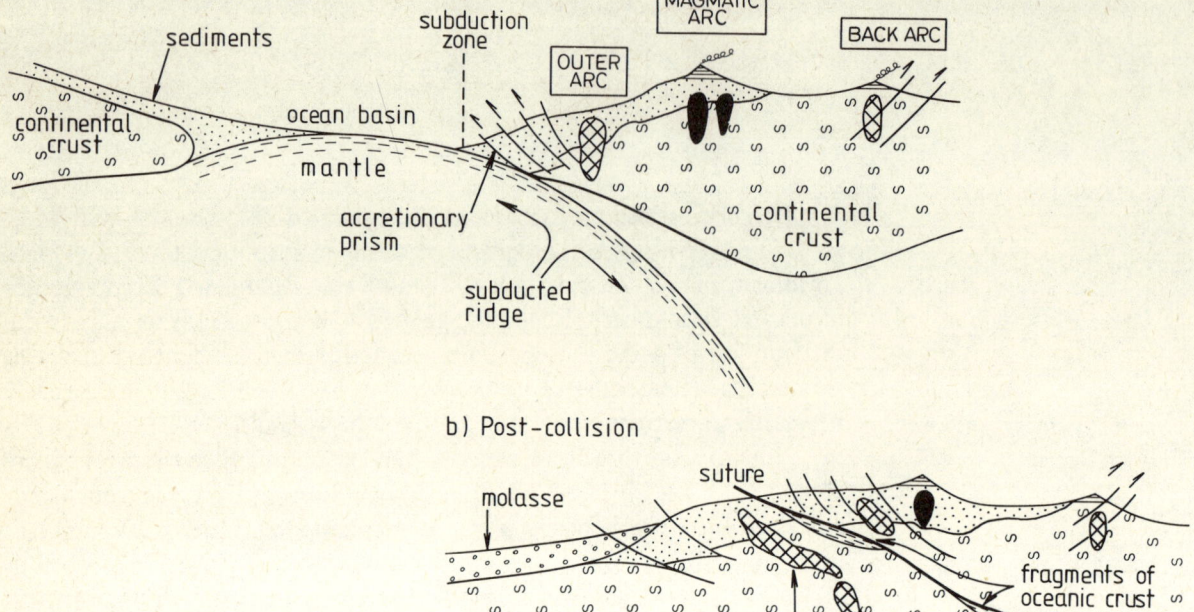

Fig. 7.1. Plate tectonic settings for magmatism and mineralisation. I-type granites black; S-type granite cross-hatched.

MAGMATIC ARCS

Magmatic arcs of cordilleran or island-arc type are the geological features most characteristic of overriding plate margins and the richest sites of mineralisation. They lie between about eighty and several hundred kilometres from the submarine trench and the underlying Benioff zone occurs at a depth of 120-180 km. The magmatic rocks are predominantly calc-alkaline in composition ranging from high-alumina basalt to rhyolite. Intrusions of dioritic or tonalitic composition occur but true granites are rare.

BACK-ARC MAGMATIC BELTS

Back-arc magmatic belts lie on the continent side of some continental margin magmatic arcs. A fold-thrust belt with thrusts dipping in the direction of the ocean may occur on the continent side of back-arc magmatic belts as in the Cretaceous fold-thrust belt of the North American Cordillera. The magmatic rocks of back-arc belts are generally more acid in composition than those of the magmatic arc and isotopic data suggests some involvement of crustal material in their petrogenesis.

Ore Deposits associated with Subduction

OUTER-ARC MAGMATIC BELTS

Outer-arc magmatic belts occur on the ocean side of some magmatic arcs. The country rock consists mainly of imbricate slices of ocean crust and flysch-type trench-fill sediments which are mechanically scraped off and emplaced tectonically above a Benioff zone. Intrusions of granitic composition associated with tin mineralisation occur in some of these outer-arc belts (for example the Shimanto Belt of south-west Japan) but closer to the ocean trench than the intrusions of the magmatic arc. Marshak & Karig (1977) have suggested that these plutons reflect the heat input provided by subduction of an ocean rise.

COLLISION BELTS

Collision belts may reflect collisions between either two continents or a continent and an island arc associated with the closure of ocean basins. The recent interest in magmatic rocks in this tectonic setting reflects increasing evidence that certain granitic magmas are generated in a subducting continental plate during collision with an overriding plate.

7.3 Mineralisation in magmatic arcs

In magmatic arcs mineralisation occurs in both intrusive rocks (e.g. porphyry copper deposits) and extrusive rocks (e.g. Kuroko-type stratiform ore bodies).

PORPHYRY DEPOSITS

Almost 100% of the world's annual production of molybdenum and over 70% of its copper is derived from 'porphyry' sources. These are large deposits, ranging from 10 to 1000 million tonnes of ore, but of generally low grade—mostly between 0.4 and 1.0% copper disseminated within the predominantly igneous host rocks. The association of the copper porphyry ores with calc-alkaline magmatism in young mountain belts, especially in the Andes and Western Cordillera, suggested that there was some relationship with subduction zones and, following the stimulus of the plate tectonic theory in the mid 1960s, exploration has found major new reserves in the Solomon Islands, Papua-New Guinea, the Philippines, Fiji, Sumatra and in the Caribbean and Mexico. These newly found deposits, and those of the circum-Pacific belt, are all associated with orogenic belts of Mesozoic and Cenozoic age and suggest that the 'porphyry' copper and molybdenum deposits are intimately connected with the subduction of oceanic crust during the last 100 million years at least. In older rocks the presence of porphyry coppers in what are now continental plates, such as in the Appalachians, the Caledonian Mountains of Scotland, the Urals, central Thailand and Taiwan, suggests that similar processes of subduction of ocean lithosphere were associated with the oceans then present prior to the collisions forming the present continental landmasses.

Both mineralised and unmineralised volcanics in island-arc environments are characterised by low initial $^{87}Sr:^{86}Sr$ ratios, generally 0.703-0.704, which are typical mantle ratios and indicate that such volcanic rocks have been derived either from the downgoing subducted slab of oceanic lithosphere or from the mantle wedge overlying the downgoing plate (Kesler *et al.*, 1975). In continental settings, especially western North America, the initial strontium isotope ratios are slightly higher, 0.704-0.709, indicating possible contamination with continental material, but the contamination involved may be small and might indicate nothing more than slight interactions between the ascending magma and the crustal rocks. Thus in both the island-arc and cordilleran orogenic belts, associated with active or previous subduction, the igneous activity and mineralisation is directly related to the mantle processes in such tectonic regimes and appears to be essentially uninfluenced by the presence of continental crust.

The primary association of porphyry deposits with calc-alkaline intrusives may be related to the partition of copper, and similar elements, between silicate magmas and the crystallising minerals (Feiss, 1978). It seems likely that the Cu^{2+} ion in a silicate melt will be preferentially partitioned into octahedral rather than tetrahedral co-ordination sites. The proportion of octahedral sites to tetrahedral sites increases with alumina content and decreases with alkali and silica contents (Burns & Fyfe, 1964). Aluminous calc-alkaline magmas would tend therefore to partition a greater proportion of copper into the residuum as crystallisation progresses, thus concentrating the copper available to be incorporated into the magmatic hydrothermal phase that eventually gives rise to the porphyry ore deposits.

The features which are typical of porphyry deposits include a plutonic or subvolcanic calc-alkaline diorite or granodiorite host rock. Pronounced hydrothermal alteration in and around this host rock which may affect an area of several square kilometres is a strongly characteristic feature of this type of deposit and is often the main clue in the field evidence leading to the discovery of porphyry deposits. The alteration usually exhibits a characteristic concentric zonation of hydrothermal mineral assemblages surrounding a central zone of potassic alteration (Guilbert & Lowell, 1974; Fig. 7.2). Isotopic and other studies (Taylor, 1974) indicate that the outer hydrothermal alteration zones are caused by convective circulation of meteoric ground waters and/or formation waters which is driven by heat loss from the cooling intrusive body. The central

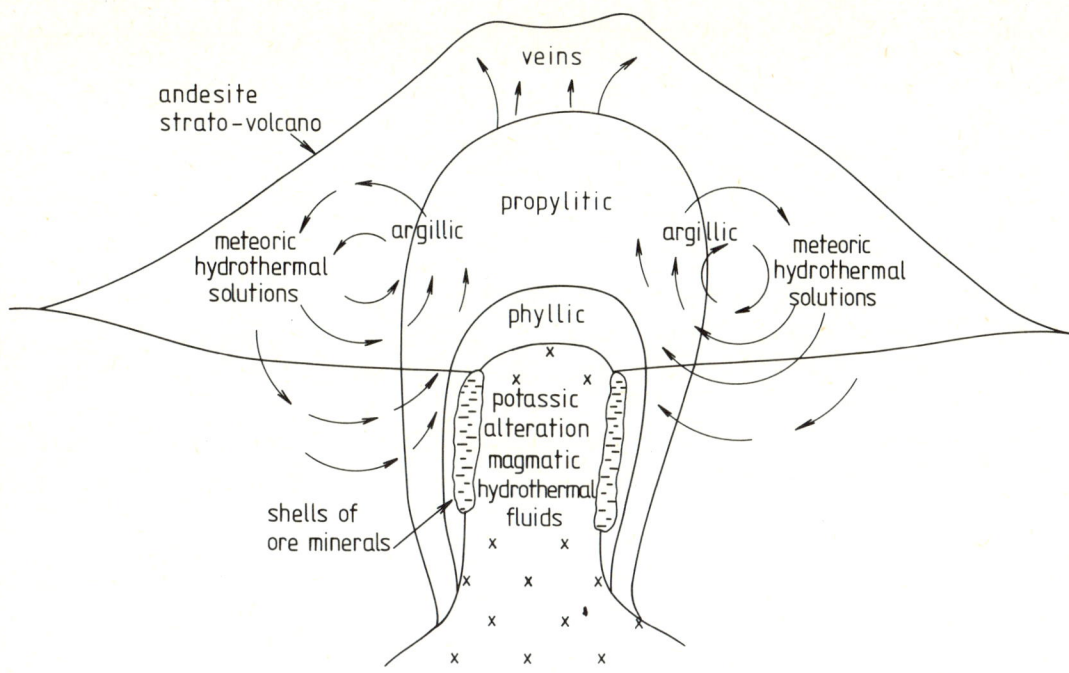

Fig. 7.2. Diagrammatic model of alteration zones around hydrothermal circulation systems in a typical porphyry copper deposit.

potassic alteration is produced by primary magmatic fluids and the main 'shells' of ore minerals tend to occur at the contact of these different fluids. The characteristic alteration zones are frequently superimposed inwards on each other because with continued cooling the hydrothermal circulation system gets smaller. The primary mineralisation usually includes the minerals chalcopyrite, bornite and pyrite. Surface oxidation and supergene enrichment (in which copper is redeposited in such minerals as chalcocite, covellite and cuprite by percolating ground waters) is an economically important process in these deposits after increasing the 'protore' grade by a factor of about five, that is up to about 1.0% copper in the supergene enrichment zone. Rich ore blankets may thus occur in supergene enrichment zones at topographic heights related to major erosional base levels.

KUROKO-TYPE DEPOSITS

Kuroko-type deposits are also characteristic of magmatic arcs but, in contrast to the huge low-grade porphyry-type of ore deposit, these are relatively small lensoid bodies containing up to a few million tonnes of ore which occur parallel to the bedding in dacite to rhyolite submarine lavas and volcaniclastic sediments (Fig. 7.3). They contain iron, zinc, lead, and copper sulphides in order of normal decreasing abundance, commonly at grades of at least 7% combined lead and zinc, and they are

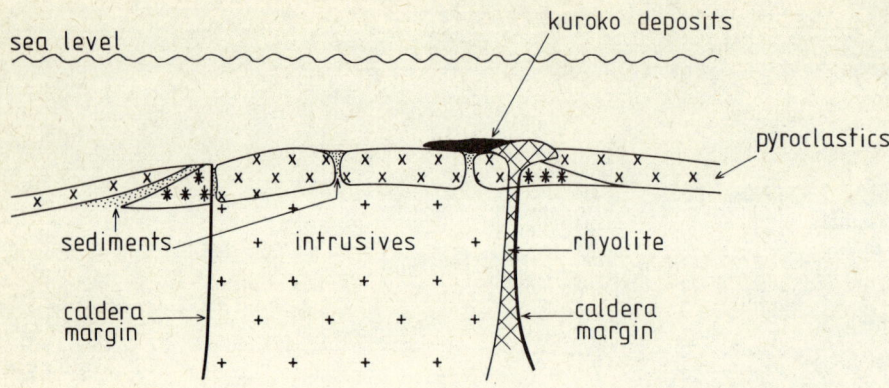

Fig. 7.3. Diagrammatic model of formation of Kuroko-type ores on the flank of a rhyolite dome in a submarine caldera.

usually worked as underground mines. Gold and silver are also recovered in small and variable amounts. Kuroko-type deposits are named from the mines in Honshu island in Japan and the best described are those at Kuroko (Horikoshi, 1969) where siliceous stockworks, sporadic gypsum deposits and disseminated ores underlie massive stratiform sulphides typically including yellow pyritic copper ore at the base overlain by black polymetallic ores carrying lead, zinc and barytes.

It is generally accepted that Kuroko-type ore bodies form on the sea floor, related to submarine hot springs in resurgent calderas or on the flanks of rhyolite domes, and that the ore fluid is a mixture of mainly sea water with a smaller component of a magmatic hydrothermal fluid. The depth of water in this ore-forming process is probably shallow because stable isotope evidence indicates that in Japan the transgressive post-Kuroko copper-lead-zinc and gold-silver veins are deposited from fluids with a large component of meteoric water which implies that the magmatic arc emerged from the sea soon after the formation of the Kuroko-type deposits.

The two types of ore deposit described above are those most characteristic of magmatic arcs. Some of the less common types of ore associated with magmatic arcs such as the magnetite-rich flows found associated with ignimbrites and other arc lavas in, for example, Chile (Park, 1961) and Pakistan (Sillitoe, 1978), or the bedded magnetite-pyrrhotite-cassiterite ores which may be of volcanogenic origin, found in the Permian magmatic arc of eastern Malaysia and Billiton, are described briefly by Mitchell & Beckinsale (1981).

7.4 Mineralisation in back-arc settings

The best example of a back-arc magmatic belt is probably the tin-silver province of Bolivia, and the so-called Western Belt of Burma and western Thailand may be a similar tectonic setting. Tin-silver production in

Ore Deposits associated with Subduction

Bolivia (Fig. 7.4) is from primary deposits associated with volcanic or sub-volcanic silicic rocks and which include complex vein assemblages and at least one large tonnage low-grade disseminated tin deposit in an igneous host rock ('a porphyry tin deposit', Sillitoe *et al.*, 1975). The Burma-west Thailand belt is a tin-tungsten province in which the mineralisation is associated only with plutonic rocks, mostly granites or adamellites. Primary mineralisation occurs as quartz vein-swarms in cusps in the roofs of plutons, disseminated deposits of Haad Som Pan type and as pegmatites and aplites intruding both granite and country rock and found especially in the roof zones of intrusions. However most of the mineral production of south-east Asia is mined from Quaternary placers derived from the primary mineralisation (see Beckinsale, 1979).

Major structural features of these back-arc magmatic belts which may be of genetic significance are that both are on the continental side of a magmatic arc and, especially in the case of Bolivia, these belts are concave towards the ocean. On its landward side the Bolivian belt is bordered by a fold-thrust belt with overthrusts, antithetic to subduction, directed towards the continent. Recent mapping in Burma indicates similar continent-directed thrusts *en échelon* on the northern border of the Burma-western Thailand fold-thrust belt. It seems possible that major movements along back-arc thrusts might result in generation of intra-continental anatectic granitic magmas very similar to those produced in

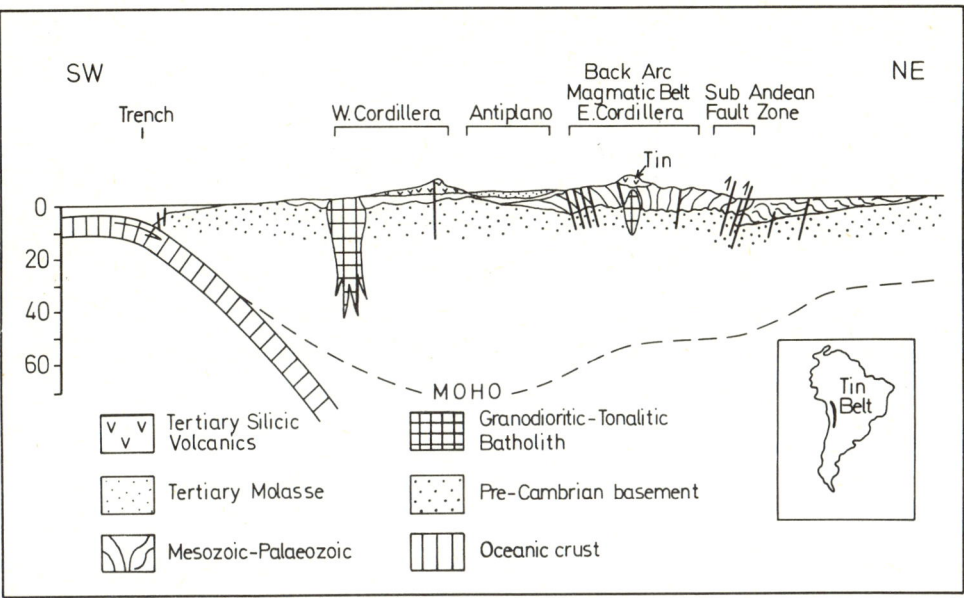

Fig. 7.4(a). South American subduction-related metallogenic zones. Cross-section of western South America showing the position of back-arc belt of Bolivia in relation to the magmatic arc of the Western Cordillera. See p. 142 for Fig. 7.4(b).

Chapter 7

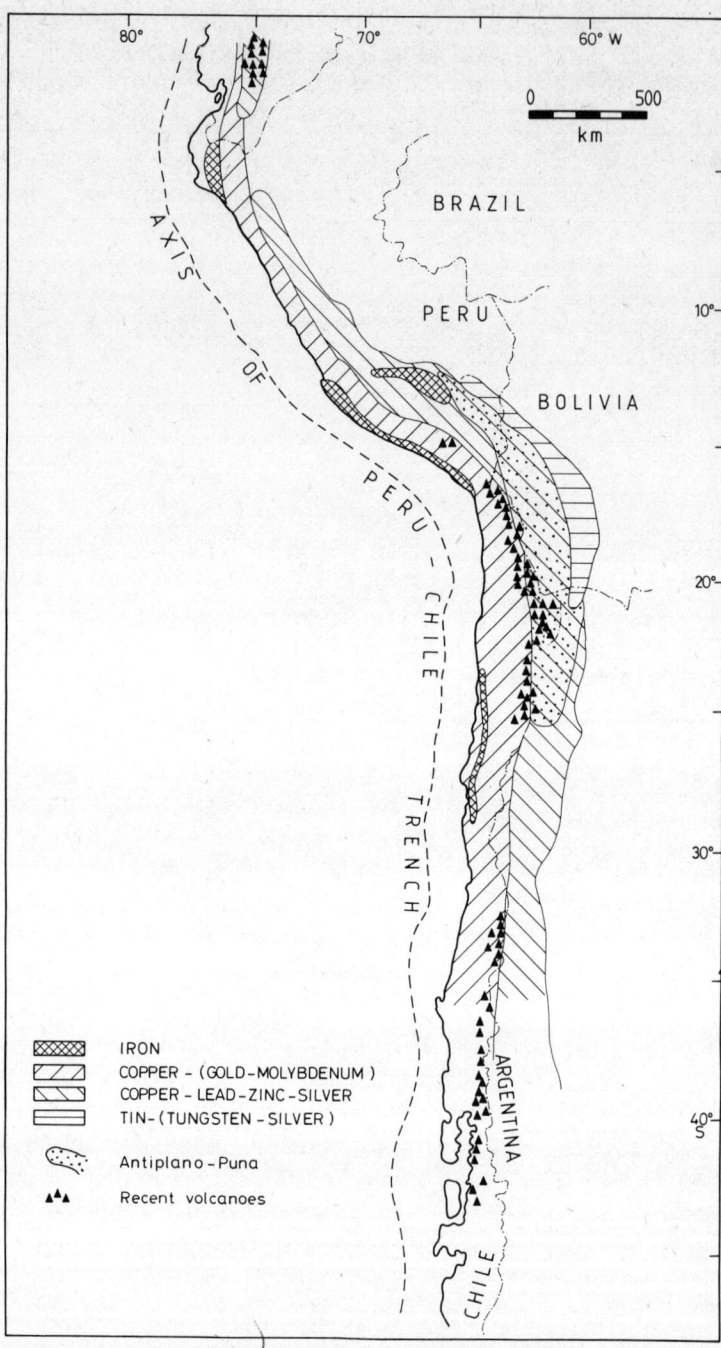

Fig. 7.4(b). South American subduction-related metallogenic zones. Map of the main metallogenic belts of western South America (after Sillitoe, 1974).

Ore Deposits associated with Subduction

collision settings (discussed below) which lead to the mineralisation described above.

Although tin-tungsten mineralisation seems to be most characteristic of back-arc magmatic belts it could be argued that some of the porphyry molybdenum deposits in Colorado also formed in a Late Cretaceous to Oligocene back-arc setting.

7.5 Mineralisation in outer-arc settings

Granites are found in the accretionary prisms of flysch-type sediments in subduction zones on the ocean side of certain volcanic arcs as discussed in the summary of different plate tectonic settings in the section above. As noted above this magmatism is thermally anomalous because the downgoing slab of oceanic lithosphere is relatively cold and Marshak & Karig's (1977) suggestion that such granites reflect heat input due to subduction of ocean rises may be a valid solution to this problem. Perhaps the best examples of mineralisation in this tectonic setting are found in Japan on Kyushu Island (Ishihara, 1978) where monzo-granite rocks support vein deposits of tin-copper-arsenic minerals associated locally with lead, zinc and fluorite.

Plutons in this setting are too close to the submarine trench to be derived from mantle melting processes on or near the Benioff zone and Ishihara (1978) has suggested that the source of both the magmas and the metals lies in the underlying continental crust. We suggest that since the trench-fill sediments are derived largely from the magmatic arc all these outer-arc granites and their associated mineral deposits reflect recycling of previously eroded components. Thus in Japan the tin-copper-arsenic association in the outer-arc magmatic belt probably mirrors the older mineralisation in the magmatic arc itself.

7.6 Mineralisation in collision belts

Collision belts appear to be characterised by anatectic granites associated with tin deposits or possibly magmatic uranium deposits. For example tin mineralisation has been described by Talalov (1977) for the margins of the Dandeldhura granite in the Lower Himalayas of West Nepal— the Himalayas being probably the best example of a collision zone. The granite exhibits extensive greisens locally and cassiterite, scheelite and tantalo-columbite occur near the margins. It should be noted, however, that although this granite could be an anatectic melt produced during the collision of the Indian plate with Asia, there is geochronological evidence that some of the Lower Himilayan granites may be older than the actual collision event (RDB unpublished).

Analogous tin-bearing granites in older orogens include the Hercynian granites of Europe and particularly the Late Triassic granites of central Thailand which extend along strike through the Main Range of Malaysia into central Sumatra (Beckinsale et al., 1979). Particular evidence which may be cited in favour of a crustal-anatectic origin for these granites in south-east Asia includes their high SiO_2 contents and restricted range of chemical compositions and extremely high initial $^{87}Sr:^{86}Sr$ ratios up to about 0.730. It is worth stressing that the tin belt of south-east Asia has yielded probably about three quarters of the total tin production of the world during the present century; south-east Asia's current annual tin production is about 230 000 tonnes. As noted above, most of the tin production in south-east Asia is mined from Quaternary placer deposits in which cassiterite has been concentrated from primary deposits by a combination of deep tropical weathering and changes in topography relative to erosional base level which has produced thick alluvial 'flats' (Beckinsale, 1979).

MAGMATIC URANIUM DEPOSITS

The Rossing deposit in Namibia is the only known major uranium ore body disseminated in plutonic rocks, and has been discussed by Berning et al. (1978) and Jacob (1978). The uranium is thought to have been concentrated during differentiation of anatectic melts and occurs mostly in the form of uraninite, disseminated in pegmatites, granites and leucogranites within high-grade metamorphic rocks of the Damara super-group. Other uraniferous granites such as those in the western Massif Central of France (Leroy, 1978; Moreau, 1977) have been compared with the Rossing deposit and the chemical composition of these granites, especially the fact that they all have high initial $^{87}Sr:^{86}Sr$ ratios suggest the possibility that they also have a collision-related origin.

7.7 Tectonic settings in relation to mineral exploration

It has been argued that the concept of plate tectonics has been of great value in mineral exploration by indicating magmatic belts favourable for particular kinds of mineral deposit. However the empirical relationship between certain mineral deposits and magmatic rocks with particular ranges of composition (for instance tin deposits with 'tin granites' or porphyry copper deposits with diorites) was used as a guide in regional exploration for many years before any plate tectonic relationships were suggested.

We believe that the geochemical characteristics of anatectic granites found in collision belts, back-arc belts and outer-arc belts are different from those of the calc-alkaline granodiorites and diorites found associated with porphyry copper deposits in magmatic arcs. The essential geochemical difference is between anatexis of crustal material leading to so called S-type granites and crustal fractionation of mantle derived parent magmas leading to the so-called I-type granite characteristic of magmatic arcs. This distinction between I and S types (Chappel & White, 1974) is a very simplistic one since in practice processes such as crustal contamination and mixing can mask features originally present in the magma. Nevertheless this granite classification may be a useful starting point in plate tectonic reconstructions: some of the main differences between I- and S-type granites are summarised in Table 7.1 (from Beckinsale, 1979) and they follow directly from the geochemical differences between a crustal and a mantle source region.

Table 7.1 Characteristic features of I- and S-type granites.

I-type or magnetite-series granites	S-type or ilmenite-series granites
Tend to be the acid end of a broad compositional spectrum from basic to acid	Tend to occur in restricted ranges of only acidic compositions
Relatively high sodium contents	Relatively low sodium contents ($<3.2\%Na_2O$ in rocks with $\sim 5\%K_2O$)
Low initial $^{87}Sr:^{86}Sr$ ratios (<0.708)	High initial $^{87}Sr:^{86}Sr$ ratios (>0.708)
Normal range of $\delta^{18}O$ values (approx. 6-10% SMOW)	Enriched in ^{18}O ($\delta^{18}O$ values $\geq$ about 10%, SMOW)
Magmas with relatively high oxygen fugacity; relatively high ferric:ferrous ratios; characterised by magnetite	Magmas with relatively low oxygen fugacity; relatively low ferric:ferrous ratios; characterised by ilmenite
Hornblende and sphene commonly present	Muscovite, monazite, cordierite and garnet commonly present

We would conclude by emphasising that it is possible that the identification of relationships between plate convergence and magma generation and metal concentration in different plate tectonic settings will be applied in future exploration. At the present time, however, it is the discovery of mineral occurrences and the nature of the igneous rocks in a magmatic belt which are the best indicators of the metals which may be present and whether the belt is mineralised.

Acknowledgement
The Director of the Institute of Geological Sciences is thanked for permission to publish this paper.

Chapter 7

References and bibliography

Beckinsale, R.D., Suensilpong, S., Nakapadungrat, S. & Walsh, J.N. (1979) Geochronology and geochemistry of granite magmatism in Thailand in relation to a plate tectonic model. *J. Geol. Soc. London,* **136,** 529-538.

Beckinsale, R.D. (1979) Granite magmatism in the Tin Belt of South-east Asia. In *'Origin of Granite Batholiths: Geochemical Evidence'* (Eds M.P. Atherton & J. Tarney) Shiva Publ., Orpington, England.

Berning, J., Cooke, R., Heimstra, S.A. & Hoffman, U. (1976) The Rossing uranium deposit, South West Africa. *Econ. Geol.,* **71,** 351-368.

Burns, R.G. & Fyfe, W.S. (1964) Site preference energy and selective uptake of transition-metal ions from a magma. *Science,* **144,** 1001-1003.

Feiss, P.G. (1978) Magmatic sources of copper in porphyry copper deposits. *Econ. Geol.,* **73,** 397-404.

Guilbert, J.M. & Lowell, I.D. (1974) Variations in zoning patterns in porphyry ore deposits. *Canad. Min. Metall. Bull.,* **67,** 99-109.

Guild, P.W. (1971) Massive sulphides versus porphyry deposits in their global tectonic settings. Print No. G 13, *MMIJ-AIME,* Joint Meeting, May 1972, Tokyo, pp. 1-12.

Horikoshi, Ei. (1969) Volcanic activity related to the formation of the Kuroko-type deposits in the Kosaka District, Japan. *Min. Dep.,* **4,** 321-345.

Ishihara, S. (1978) Metallogenesis in the Japanese island arc system. *J. Geol. Soc. Lond.,* **135,** 389-406.

Jacob, R.E. (1978) Granite genesis and associated mineralisation in part of the Central Damara Belt. In *Mineralisation in metamorphic terrains* (Ed. W.J. Verwoerd) van Schaik, Pretoria, pp. 417-432.

Kesler, S.E., Jones, L.M. & Walker, R.L. (1975) Intrusive rocks associated with porphyry copper mineralisation in island arc areas. *Econ. Geol.,* **70,** 515-526.

Marshak, R.S. & Karig, D.E. (1977) Triple junctions as a cause for anomalously near-trench igneous activity between the trench and volcanic arc. *Geology,* **5,** 233-236.

Mitchell, A.H.G. (1974) Southwest England granites: magmatism and tin mineralisation in a post-collision tectonic setting. *Instn. Min. Metall. Trans. (Sect. B: Appl. Earth Sci.),* **83,** B95-97.

Mitchell, A.H.G. (1977) Tectonic settings for emplacement of southeast Asian tin granites. *Geol. Soc. Malaysia Bull.,* **9,** 123-140.

Mitchell, A.H.G. (1978) Geosynclinal and plate tectonic hypotheses: significance of late orogenic Himalayan tin granites and continental collision. *11th Commonwealth Min. Metall. Cong., Instn. Min. Metall.,* Hong Kong, Paper 37, pp. 1-13.

Mitchell, A.H.G. (1979) Rift, subduction and collision-related tin belts. *Geol. Soc. Malaysia Bull.,* **11,** 84-102.

Mitchell, A.H.G. & Garson, M.S. (1972) Relationship of porphyry copper and circum-Pacific tin deposits to palaeo-Benioff zones. *Inst. Min. Metall. Trans. (Sect. B: Appl. Earth Sci.),* **81,** B10-B25.

Mitchell, A.H.G. & Garson, M.S: (1976) Mineralisation at plate boundaries. *Mins. Sci. Eng.,* **8,** 129-169.

Park, C.F. (1961) A magnetite flow in northern Chile. *Econ. Geol.,* **56,** 431-436.

Sillitoe, R.H. (1970) South American porphyry copper deposits and the new global tectonics. *Resumenes Primer Congr. Latinoamericano Geol.* Lima, Peru, pp. 254-256.

Sillitoe, R.H. (1974) Tectonic segmentation of the Andes: implications for magmatism and metallogeny. *Nature,* **250,** 542-545.

Sillitoe, R.H., Halls, C. & Grant, N.J. (1975) Porphyry tin deposits in Bolivia. *Econ. Geol.,* **70,** 913-927.

Sillitoe, R.H. (1978) Metallogenic evolution of a collisional mountain belt in Pakistan: a preliminary analysis. *J. Geol. Soc. Lond.,* **135,** 377-387.

Snelgrove, A.K. (1971) Metallogeny and the new global tectonics. *Mineral Res. and Explor. Inst. Bull.,* No. 76, Turkey, pp. 130-149.

Talalov, V.A. (1976) Main features of magmatism and metallogeny of the Nepalese Himalayas. In *Colloque sur l'écologie et la géologie de l'Himalaya.* Centre National de la Récherche Scientifique No. 268, Paris, pp. 409-430.

Taylor, H.P. (1974) The application of oxygen and hydrogen isotope studies to problems of hydrothermal alteration and ore deposition. *Econ. Geol.,* **69,** 843-883.

Walker, R.R., Matulich, A., Amos, A.C. Watkins, J.J. & Mannard, G.W. (1975) The Geology of the Kidd Creek Mine. *Econ. Geol.,* **70,** 80-89.

Chapter 8

The Origins of Ore Deposits in Sedimentary Rocks

J.P.N. BADHAM

8.1 Introduction

The ore deposits considered in this chapter are all in sedimentary rocks and have originated from relatively normal sedimentary or diagenetic processes. Thus for an understanding of the origin of mineralisation it is imperative that the origins of host rocks are also clearly understood. Ore, as the term is currently used, refers to mineralisation that can be mined profitably. This clearly changes with changing economics and technologies so the following sections, while dealing with major ore types currently exploited, will be essentially concerned with mineralising processes rather than with absolute tonnages and grades.

Naturally not all sedimentary ores, nor even all examples of those described, can be discussed in this chapter. Some specific ore types have been omitted including some residual ores (gossans and laterites), ironstones, manganese formations and uraniferous black shales. It is thought, however, that the principles outlined in connection with the deposits that are described here can be relatively easily applied to these other deposits. The emphasis has been given to deposits that seem to have a distinct relationship with geotectonic processes.

Just as there are both clastic and chemical sediments, there are clastic and chemical ores. Just as post-depositional processes may substantially alter the nature of primary sediments so may they also form secondary ore deposits. As long as sediments retain or develop porosity and permeability they remain potential receptors for mineralising fluids. There are of course no clear divisions between primary and secondary processes but the following broad definitions are adopted here.

Primary—the products of sedimentary processes at or above the sediment-water interface. They may be clastic, chemical or both.

Secondary diagenetic—the products of interactions between primary sediments and pore waters derived essentially from their dewatering.

Secondary epigenetic—the products of interactions between sedimentary rocks and pore waters which attained their physical and chemical conditions elsewhere. The term in no way requires any connection with igneous processes but equally in no way precludes it. Only deposits for which there is no clear igneous source will be considered here.

Chapter 8

Primary sedimentary ore deposits may occur in specific geotectonic settings but are essentially controlled by the physical and chemical conditions of sedimentation: climate and palaeogeography may therefore be particularly important controls. Secondary sedimentary deposits are more closely related to basin evolution and fluid migration: they may be more closely controlled by geotectonic processes, particularly in as much as heat and rates and routes of migration are important. Thus both geotectonic and environmental factors may be important. Most of the deposits described are the products of relatively simple chemical or physical processes which may occur in many different geotectonic and environmental settings. What is important for ore genesis is the significant enhancement of these processes. It is important to separate the genetic processes which may produce mineralisation from the various factors which may enhance them and produce major ore bodies.

Many of the deposits described are Precambrian. While the geotectonic processes of this period are becoming better understood not all of them can be related to plate tectonics as seen today. With changing patterns of geotectonics different enhancement controls may have operated at different times to produce grossly similar mineralisations. Conversely some processes may be unique in time and so may be the resultant ores. Similarly, and with equal effect, environmental conditions may have changed. For the purposes of this chapter Cloud's (1976) fourfold division of Earth history is adopted.

The *Archaean* (Permobile regime of Windley, 1977) covers the period from the earliest dated rocks to a period between 3.0 and 2.3 billion years. A mobile tectonic regime is indicated by high-grade metamorphic, granitic and greenstone terrains providing little preservation potential for sedimentary rocks and the environments which they represented. Greenstone belts contain the only well-preserved sedimentary sequences and these are dominated by volcanoclastics and greywackes. Sediments that may be confidently ascribed to stable shelf or continental areas are rare. The *Proterophytic* (Lower Proterozoic) dates from the end of the Archaean to about 1.8 billion years—a time of worldwide orogenic culmination. The period is characterised by the first mature sedimentation on stable shelves and huge outpourings of basic lavas. Some of the culminating orogenic belts appear closely similar to Phanerozoic ones and have been accorded a plate tectonic origin (Hoffman, 1973; Badham, 1973, a,b). Others clearly evolved by different means (Dimroth, 1972; Kröner, 1977). The *Palaeophytic* (Middle and Upper Proterozoic) is characterised by vertical tectonics and graben formation in the interior of large continental plates on which widespread epicontinental areas developed. Plate tectonic processes were certainly active and came to

Ore Deposits in Sediments

dominate Earth processes in the *Phanerozoic* with the disruption of the larger continental plates. In many ways the distinction between Phanerozoic and Proterozoic is arbitrary and may depend on the absolute size of continental plates at any time.

Cloud (1976) set up his divisions to document stages in the development of atmosphere, hydrosphere and biosphere. Since sedimentary rocks are the products of interactions between these and the lithosphere the nature of sediments—and thence of associated ore deposits—may have changed. The biosphere has clearly changed with time, although whether the biomass has is a more contentious issue. There is active debate over the nature and extent of change in atmosphere and hydrosphere. Many workers (Cloud, 1972; Roscoe, 1973; Schopf, 1978) believe the atmosphere to have been anoxygenic prior to 2.3 billion years. Others, however (Dimroth & Kimberley, 1976; Towe, 1978; Clemmey & Badham. In prep.) argue that it has always contained free oxygen, but that the $O_2:CO_2$ ratio may have changed to a greater or lesser extent. The details of the debate are not germane here but the fact that the nature and abundance of various types of sedimentary ore deposit may have changed with time certainly is.

8.2 Primary sedimentary deposits

PLACER DEPOSITS

For a mineral to accumulate to ore grade during normal clastic sedimentary processes it must be stable during weathering and transport, and must be dense. Minerals commonly occurring in placers therefore include oxides (magnetite, rutile, cassiterite) and native metals (gold, platinum, diamond). The formation of clastic sediments involves weathering, transport, deposition and preservation: for ore formation any or all of these must act to concentrate selectively the ore mineral. Common resistate minerals may require no special source and may occur in varied clastic environments, but the rarer minerals are unlikely to reach ore grade as placers unless source areas were significantly enriched (possibly even to ore grade) themselves.

Selective concentration of resistate minerals may occur during weathering in humid, tropical conditions where other minerals in the rock are chemically destroyed. Thus climate may play an important part. During transport dense resistate minerals may be continually sorted from other clastic material. In particular accumulations may occur on point bars in meandering rivers and longitudinal bars and abandoned channels in braided rivers. Preservation of such sequences as terraces by channel migration or rejuvenation may locally form ore, e.g. the diamondiferous river gravels of Botswana or the auriferous gravels of the Klondike (Hester, 1970). However, the preservation potential of fluviatile gravels is low and most economic deposits of this type are less than 10 million years old.

Fluviatile transport may involve continual reworking leading to further concentration and the process may continue where rivers enter the sea, particularly where the deltas are wave dominated. Continual sorting and reworking on beaches by a combination of wave action, longshore drift and wind has produced the titaniferous oxide ores of Florida and northeast Australia—and indeed many other concentrations which are not economic at present. However these processes alone appear generally to have been insufficient to produce large tonnages of rare metals such as gold. For these it would seem that a number of periods of erosion and deposition with repeated concentration are necessary—and these conditions imply intermittent uplift in source areas and thus tectonic activity.

In general there would appear to be few geotectonic controls of placer formation which occurs in rivers and coastlines on both stable and unstable margins. However, the very large concentration factors needed for the rarer minerals require enriched source areas and repeated reworking during transport. Source areas may be earlier placer concentrations but will mainly be magmatically active terrains where primary enrichments due to igneous and metamorphic processes have taken place. As should be evident from Chapter 7, such enrichments characterise particular geotectonic environments: as a consequence so too will the associated placer deposits. The Kinta Valley tin deposits, Malaya, (Dixon, 1979) exist only because of the previous concentration of tin in veins and skarns around granites, themselves the product of subduction (Mitchell & Garson, 1973). Intermittent but frequent reworking may occur during periods of progressively lower sea-level related to glaciation, but is more general in areas of tectonic instability: again active as opposed to passive continental margins are favoured. Henley & Adams (1979) note the requirement of two complex stages for formation of 'giant' gold placers—initial enrichment in the source area and then repeated mechanical concentration in successive generations of alluvium. They identify particularly the circum-Pacific orogens as ideal with the change in plate motions throughout the Mesozoic and Cenozoic causing continual emergence of older auriferous orogenic belts and continual reworking in derivative clastic sediments (Fig. 8.1). The placers in California, the Yukon, Alaska, British Columbia, New Zealand and Siberia all fit this model nicely; the relationship between ore and geotectonics is evident.

It is implicit from these descriptions that most placer deposits are relatively young. While there are older examples they are generally small

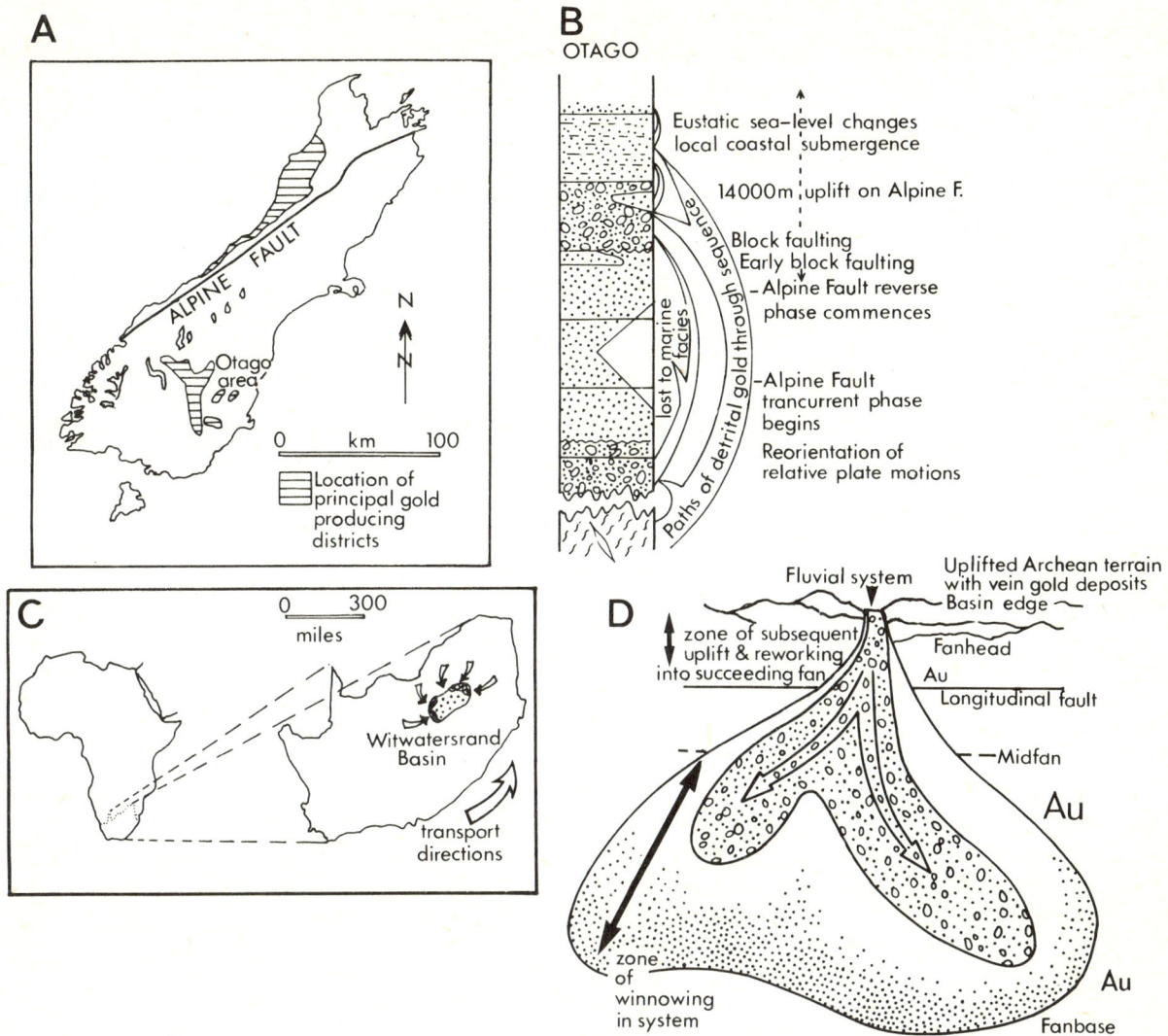

Fig. 8.1. Ancient and modern placer gold deposits. A & B. The Tertiary-Recent deposits of New Zealand showing the complex interplay of intermittent uplift and continual reworking of alluvium (modified from Henley & Adams, 1979). C & D. The Lower Proterozoic deposits of the Witwatersrand showing the relationship between basin margin faulting, sedimentation and reworking (modified from Pretorius, 1975).

and the application of any understanding of a relationship between plate tectonics and placer deposition is hardly likely to help in the location of ore: usually subsequent tectonic processes have obscured the relationships. Henley & Adams note the occurrence of auriferous conglomerates in Appalachian molasse, but propose no sensible geotectonic synthesis for their origin. Schneider & Lehmann (1976) suggest that much of the magmatic tin in Bolivia may have been recycled palingenetically from Silurian placers, but again offer no geotectonic setting for the placers. It would appear therefore that the preservation potential for placers, especially those in the circum-Pacific region, is low and an obvious

Chapter 8

deduction would be that there should be fewer placer deposits in progressively older rocks. This generalisation is valid for rocks younger than 1.8 billion years but is manifestly not so for Lower Proterozoic strata.

The lowermost Proterozoic strata in almost all the shield areas of the world are clastic fluviatile to shallow marine sequences which lie unconformably on Archaean basement. Virtually everywhere they are mineralised with gold and uranium: the total tonnage of gold in the Witwatersrand and Jacobina systems is an order of magnitude more than that in all the Tertiary-Recent placers described above. There is some argument as to whether or not the associated uranium—especially enriched in the Huronian Supergroup—is of detrital or secondary origin: a detrital origin would require an anoxygenic atmosphere for uranium is highly soluble in oxidising conditions. This debate is not relevant here, but the sheer volume of gold mineralisation, which is independent of the oxidation state of the atmosphere, certainly is. Clearly some or all of the factors promoting placer formation must have been enhanced. The source areas, Archaean greenstone and granite terrains, would have been (and still are in many places) particularly rich in gold. The gold now occurs in mineralogically mature quartz pebble conglomerates indicating chemical weathering and preconcentration of resistates in the source areas. The deposits show evidence of frequent reworking and further sorting in rivers, fan heads and beaches (Pretorius, 1975) (Fig. 8.1). Indeed all the factors enhancing placer gold formation appear to have been better in the Lower Proterozoic than subsequently. In addition the preservation for the first time in Earth history of thick sequences of clastic sediments at the interface between stabilised and unstable continental crust promoted the growth of huge deposits. The fact that the depositional basins were not subsequently involved in orogenesis has led to their preservation—these deposits are a unique product of a unique set of geotectonic conditions.

CHEMICAL SEDIMENTS—IRON FORMATIONS

Most elements may be precipitated as primary sediments from surface waters but usually either the conditions for precipitation are so extreme as not to occur in the natural environment or the surface waters are so dilute that ore-grade concentrations are not reached. Oxides, carbonates, silicates, sulphates and sulphides are all known and examples of both inorganic and organic precipitation are well documented. In contrast chemical precipitates may form from metal solutions discharged subaqueously—such solutions may be magmatically derived and/or driven, but their deposits are considered to be volcanogenic and are described

Ore Deposits in Sediments

elsewhere (Chapters 6 & 7). However non-magmatic fluids could also be discharged and precipitate their solutes. Because the precipitation controls in all these cases are those occurring in the surface environment the changing conditions during burial and diagenesis may substantially affect the deposits. It is a moot point whether some deposits owe their enrichment to ore grade to primary or secondary sedimentary processes.

Only iron formations are considered in this section. Of the other possible examples consideration of lead-zinc deposits in shales will be deferred until the last section for reasons that will become apparent.

James (1954) defined Iron Formation as a chemical sediment containing 15% or more iron of sedimentary origin. Iron formations are known throughout the geologic record, indeed they occur in the oldest rocks known, the Isua supracrustals of Greenland. However, some care must be taken to distinguish iron formations of volcanic origin (Algoma-type) which are particularly abundant in Archaean terrains and which occur frequently associated with volcanogenic mineralisation in younger terrains. Despite being a relatively common lithology by far and away the greatest volume of iron formation was deposited during the latter parts of the Lower Proterozoic. In particular a facies of iron formation with iron oxides thinly but persistently banded with chert (Banded Iron Formation or BIF) provides the major part of world iron production today.

Major iron formations occur on all the world's cratons (Fig. 8.2) in a variety of facies associations. In general they lie unconformably on peneplained Archaean cratons or the lowermost Proterozoic clastic sequences that so commonly contain placer gold. The huge basins in

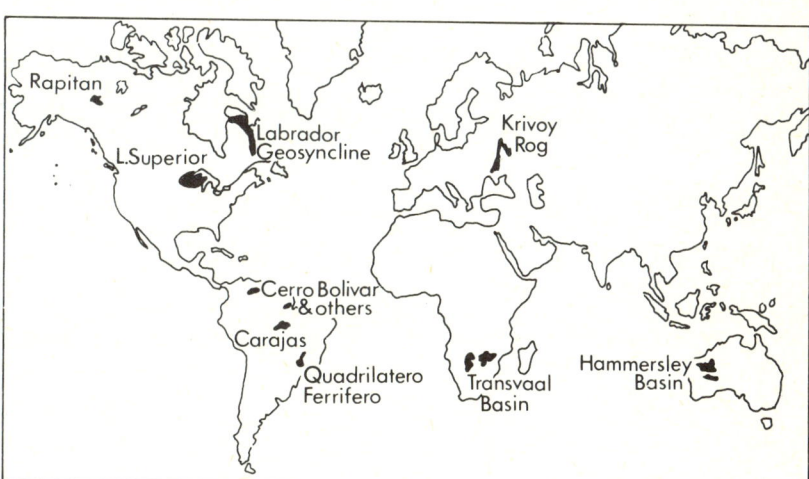

Fig. 8.2. Distribution of the major iron formations of the world.

Chapter 8

which they formed (e.g. 400 000 km² for the Transvaal Basin: Button, 1976a) were gently subsiding areas of continental crust. Initial subsidence was on basin margin faults and large piles of clastic mafic volcanic rocks were developed. An extensive chemical sedimentary unit, predominantly carbonate, but containing the iron formations, overlies the basal unit. In many cases an upper clastic unit overlies this usually with some disconformity (Fig. 8.3).

Area	Hammersley[1] Basin	Transvaal[1] Basin	Marquette[2] Lake Superior	Krivoy[3] Rog	Quad. Ferrifero[4] Brazil
Upper Clastic Unit	Mount Bruce Supergroup Wyloo Group	Transvaal Supergroup Pretoria Group	Baraga Group	Upper K_3	Itacolomi Series Minas Series (Piracicaba Fm.)
Chemical Sedimentary Unit	Hammersley Group	Chuniespoort Group	Menominee Group	Middle K_2	Minas Series (Itabira Fm.)
Basal Volcanic and Clastic Unit	Fortescue Group	Wolkberg Group	Chocolay Group	Lower K_1 / Metabasite	Minas Series (Caraca Fm.) (Tamandua) Fm.
Early Proterozoic Clastic Unit	Minor	Witwatersrand Supergroup	Huronian Supergroup	?	Jacobina Series
Archaean Basement	Granite–Greenstone	Granite–Greenstone	High-grade terrain	Migmatites	Granite–Greenstone

Fig. 8.3. Stratigraphic columns of some of the major iron formations showing the extraordinary similarity of their evolution in widely separated locations. Columns are simplified from data in: Button, 1976a(1); Bayley & James, 1973(2); Alexandrov, 1973(3); and Dorr, 1973(4).

There is widespread agreement (Button, 1976b) that the chemical sedimentary unit was deposited in very gently sloping basins fringed by deeply weathered, peneplained terrains which provided no clastic debris. The periods of ferruginous sedimentation correspond to periods of restricted circulation behind an offshore bar which may have been tectonic (James, 1954) or sedimentary and biogenic (Button, 1976b). The iron formations are always arranged in a set sequence of facies (oxide-carbonate-sulphide, with silicates overlapping all three zones) going away from the shoreline. This has been used by many workers to suggest that the ferruginous sediments are entirely primary. Despite a number of intricate models for the supply and precipitation of such huge volumes of iron, no consensus has been reached and these remain the principal problems for primary genesis theories. Button (1976a and b),

Kimberley (1974) and Dimroth (1977a and b) have all proposed that iron formations are essentially the product of diagenetic replacements of primary carbonates. There is clear petrographic and chemical support for significant replacement and enrichment of primary carbonate, but again this theory fails to account for the vast amount of iron present in this unique time-facies association.

Iron is essentially insoluble in the trivalent but highly soluble in the reduced, divalent state. The natural environment is normally slightly oxidising and mildly alkaline. Under these conditions ferric species are stable. Consequently the supply of large volumes of iron to the sedimentary environment presents a problem for which four types of solution have been proposed. Borchert (1960) suggested that ferrous iron could be mobilised from clastic sediments by reduced diagenetic fluids either onto the sea floor or into oxidised strata, there to precipitate ferric minerals. The process was documented on many modern deltas and the different iron minerals were found in the observed facies sequence. However the process is unable to explain the volumes of iron formation or their unique time-facies restriction. Button (1976a) suggested that the iron was enriched in restricted basin brines by evaporation. He envisaged the iron being derived by ground water flushing into the basins from deeply weathered profiles on the surrounding land areas. Holland (1973) noted that iron may also be supplied from the seaward side by upwelling iron-rich deep basin waters. Both these models are seen to apply to Phanerozoic ironstones, but cannot explain the huge volumes of iron in the older ones. Nor does either theory attempt to account for the huge size of the basins but it is implicit in both that if, for whatever reason, the Lower Proterozoic basins were bigger then the iron supply and precipitation would have been correspondingly greater.

Many workers have tried to solve the iron supply problem by suggesting that solutions of reduced, acid, ferruginous brines were supplied to the basin by volcanic activity. However the clear lack of any directly associated volcanism with these iron formations and their facies restriction makes this most unlikely. It is equally unlikely that the solutions escaped through the sedimentary pile from the residually warm basal volcanic unit: in many cases these volcanics were sub-aerial and the chemical sedimentary unit overlies them with some unconformity. It is probable that the volcanic piles had cooled prior to iron deposition and could not have supplied warm ferruginous solutions.

Cloud (1972) proposed a resolution to the paradox as part of his overall model of atmospheric evolution. He contends that prior to atmospheric oxidation continental weathering would have removed all iron, as ferrous species, to the sea which thus became saturated with ferrous carbonate. Oxygen-producing organisms sequestered the iron to precipitate ferric oxides, especially in seasonal bands, explaining the finely banded nature of oxide-facies BIF. Once all the iron had been precipitated by rapid diversification of these organisms there would be no further chemical supply by weathering which had consequently become oxidative. This elegant theory certainly explains the abundance of Lower Proterozoic iron formation but must be treated with caution. Quite apart from telling geological evidence for free oxygen in the atmosphere prior to the Lower Proterozoic, there is no indication that Archaean weathering profiles are any more or less iron depleted than modern ones. Equally there is no geological evidence at all that Archaean seas were saturated with ferrous carbonate, and fossils of Archaean photosynthetic organisms are not mantled by oxidised iron species. These objections have been partially resolved by considering the Archaean and Lower Proterozoic atmospheres as having had free oxygen but with a lower $O_2:CO_2$ ratio than today's. The enhanced acidity and reduced Eh would encourage transport of ferrous species during weathering (Drever, 1974).

However the problem of relationship between atmosphere and iron formation is exacerbated by the huge Rapitan iron formation in Upper Proterozoic strata in north-western Canada. Morphologically and in tectonic setting this iron formation is similar to the older ones but here there is a close association with glaciogenic strata (Young, 1976). Yeo (1978) proposes a model involving mixing of cold, sub-ice shelf waters with iron- and silica-enriched waters from a nearby spreading centre. Whatever the origin of these rocks it implies that atmosphere was of less importance than setting.

A combination of different atmospheric conditions, of both primary and diagenetic iron precipitation and of geotectonic conditions favouring widespread epicontinental seas with subdued tectonically stable hinterlands, would seem to provide the best explanation for these enigmatic deposits. That a critical combination of geotectonic conditions was required is shown by the frequent occurrence of iron formations at the same interval of continental evolution but at different actual ages in different places.

The average grade of primary iron formations is usually between 20 and 40 wt% Fe. In general this is too low a grade to mine profitably and most new operations would require grades in excess of 60%. That many of these deposits have such grades is due to post-depositional oxidative weathering having removed silica and carbonate. To have been exposed and weathered, and then covered and preserved again over such vast areas, these enriched iron formations could only have survived

Ore Deposits in Sediments

on cratonic areas which have remained stable since iron formation deposition. This provides a further geotectonic constraint on the location of iron formations. It also perhaps explains why Phanerozoic deposits are so small for with the advent of plate tectonics shelf sedimentary sequences became more restricted in time and space and were more frequently disrupted during orogeny. Widespread long-lasting epicontinental marine deposits on the scale of those in the Lower Proterozoic have not been developed subsequently. In many ways the modern analogue of much of these iron formations may be the manganese-iron rich sediments and nodules of the deep ocean floor—an environment selectively destroyed by plate tectonic processes.

8.3 Secondary sedimentary deposits

INTRODUCTION
Secondary sedimentary deposits involve the migration of pore waters through and possibly out of a lithifying sedimentary pile and localised precipitation of their solutes. Problems in genesis of these deposits involve the source of metals, the locus of precipitation and the nature of the precipitant. The greater part of such deposits is of base metal sulphides so we must ask what special conditions cause enrichment of base metals in normal diagenetic fluids and what special conditions allow their reduction and/or reaction with sulphur. Base metals are particularly enriched in volcanic strata and argillaceous sediments whence they are relatively easily removed in solution: chloride complexing has been shown (Helgeson, 1970) to be a particularly efficient way of keeping metals in solution, and these complexes are stable in mildly reducing, mildly acidic conditions. Such conditions pertain during the de-watering of marine pelitic sediments where sea water provides an unlimited supply of chlorine.

The base metals have increasing solubilities as sulphides and stabilities as chloride complexes in the order iron-copper-zinc-lead. We might expect polymetallic deposits formed by precipitation from pore waters to be zoned. However selective fixing of conditions at either source or deposition site could cause selective transport and depostion of only one of the base metals leading to essentially monomineralic deposits. Similarly if the source rocks happen, for whatever reason, to be enriched in only one of the elements, then only deposits of that element are likely to form.

The main precipitant for the base metal chloride complexes is the sulphide ion. This could be present in the ore solution, which would then only be stable at relatively high temperatures ($>150°C$). It could be reduced from sulphate at the site of deposition—and sulphate itself has an abundance of potential sources. It could also be already present as H₂S.

There is an enormous variety of sedimentary prisms that could feed such mineralising systems. Not only can the source rocks vary, but so can their thermal evolution causing fluids of different chemistries and temperatures to migrate at different times and different rates. The deposition sites and precipitation mechanisms may be equally varied and one might reasonably expect a bewildering variety of deposits. The fact there are relatively few types of any significance suggests that a whole series of critical conditions have to be met coincidentally to make ore (Fig. 8.4). In the following sections the three major types of mineralisation are described. Note that there are deposits of each type that range from essentially syngenetic to thoroughly epigenetic—in the way chosen to use the word. Not surprisingly there are widespread differences of opinion concerning the origins of these various deposits.

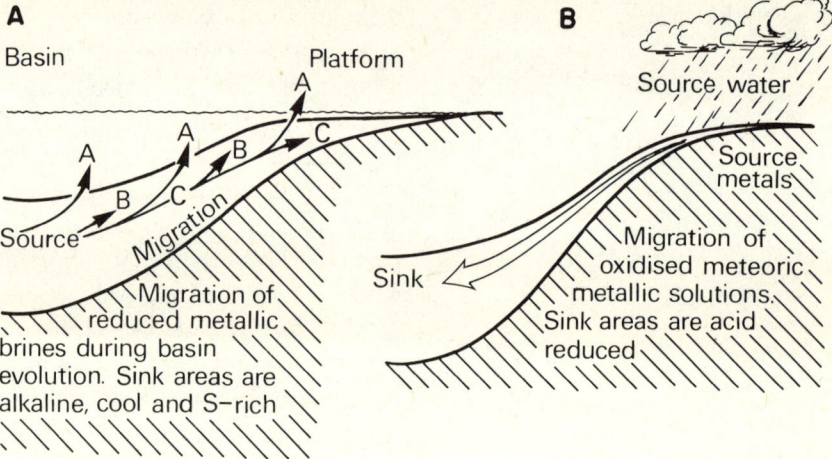

Fig. 8.4. Schematic diagrams showing the two possible sources of migration paths of diagenetic fluids in depositional basins.
A Dewatering during compaction into syngenetic (A), diagenetic or unlithified (B) and epigenetic or lithified (C) environments.
B Descendant migration of surface water into porous sediment pile after deposition. Model (a) may apply to metal sources for red-bed copper and the various Zn-Pb deposits. Model (b) may apply to S sources for the metals in (a) and to Colorado Plateau-type U-V-Cu deposits.

THE RED BED–COPPER ASSOCIATION

Although this title is something of a misnomer because the copper rarely actually occurs in the red beds, red-stained continental arenites are a ubiquitous associate of this style of mineralisation. The deposits discussed included the Kupferschiefer, the Zambian Copperbelt, the White Pine Copper district, the Angolan Copper district and some recent deposits in the Atacama desert. The deposits range in age from Middle Proterozoic to Recent. Lower Proterozoic examples are known but none are mined as yet. The ore type shows no particular distribution with time, but the geotectonic processes of the Middle and Upper Proterozoic seem to have encouraged greater size.

Mineralisation is uniquely confined to the first marine sediments deposited under an arid climate on a highly weathered basement and overlying red continental clastic sediments. The metalliferous strata are frequently overlain by a sequence of carbonates and evaporites. The transgressive sequence may have occurred in response to rifting and sagging of continental crust and the underlying strata may contain greater or lesser volumes of mafic volcanics. The mineralisation may be enormously widespread although economically recoverable sections are generally limited—both primary variations and post-depositional effects may play a part in localising ore. The mineralisation is commonly zoned outward from the old shoreline from copper to zinc/lead and upwards in the same sense. The mineralisation is generally stratiform and stratabound but in detail transects facies and cross-cuts bedding. Organic-rich shales and sandstones are particularly enriched in copper and intervening (usually algal) carbonate beds are often barren. There is clear evidence that much of the mineralisation in sandstones was secondary, precipitated on reduced shale clasts or on earlier pyrite. The petrographic evidence from shales is more equivocal. There is clear evidence from textures, from associated organics and from isotopic studies that temperatures involved in these mineralisations never exceeded 100°C. Consequently it is improbable that sulphur and the metals travelled in the same solutions. (In some instances there is clear evidence for accumulation of detrital copper sulphides in arenites, but rarely to any significant grade.)

In the Central African (Zaire and Zambia) Copperbelt there is widespread copper (and cobalt) mineralisation in Upper Proterozoic sediments of the Katangan system (Fig. 8.5). There is a basal sequence of continental clastic strata giving way firstly to interfingering carbonaceous shales and wackes with algal carbonates and then to carbonates with evaporites. The sequence overlies a basement complex which had been substantially weathered but on which there was considerable relief. Transgressive sequences are locally developed in a number of cycles. Copper mineralisation is most abundant in the shales and wackes, but both diagenetic mottling of underlying sandstone and replacement of shale clasts in arenites by copper minerals is commonplace.

Geotectonically the Central African province was part of a huge subsiding basin with clastic strata predominating in Zambia and carbonates in Zaire (Mendelsohn, 1961). The Copperbelt formed on the edges of this basin where differential vertical movements were most pronounced (Brock, 1961). Brock suggested that the main depression was a rift, but the base of the system is not seen, the few igneous rocks are gabbros that occur high in the sequence (Mendelsohn, 1961) and there is little

Chapter 8

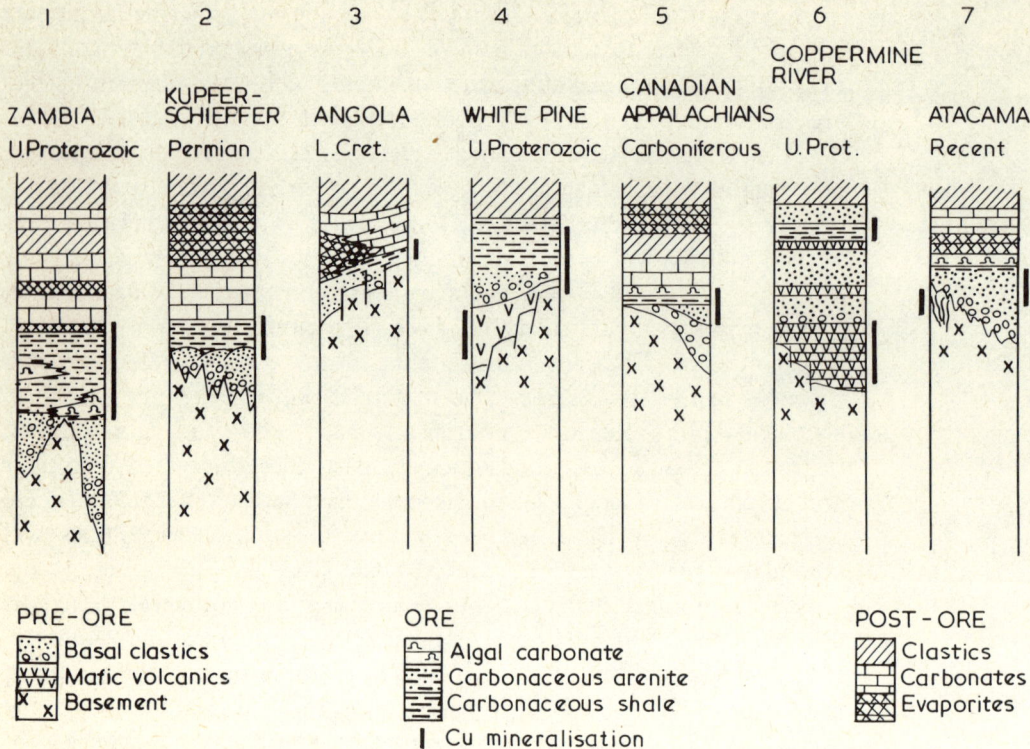

Fig. 8.5. Stratigraphic columns for red-bed copper mineralisations, generalised from references cited in the text. Column 7 is hypothetical—the possible product of transgression of the saline lagoonal facies of the Boca de Vitrila over the sediments of the Atacama desert.

direct evidence of rifting. Sawkins (1976) also espouses the rift hypothesis but offers no more evidence than that in Mendelsohn's book.

The Kupferschiefer ores of Europe are remarkably similar in style and setting to those in Zambia. The ore-bearing strata are again carbonaceous shales lying above continental clastic sediments and beneath carbonates and evaporites (Fig. 8.5). The strata were deposited in front of and from the Permian Zechstein sea that transgressed south and eastwards from an open marine connection near what is now the northern North Sea (see Russell & Smythe, 1977). The mineralised shales extend from England to Poland but are rarely of ore grade. The underlying strata are variable in thickness and higher grade mineralisation overlies thicker footwall sequences. The underlying strata are also diagenetically reddened where the mineralisation is richer. Mineralisation is zoned upwards and laterally from copper to zinc to lead. There is sparse mineralisation in footwall and hanging wall strata.

The tectonic setting of the Zechstein is of widespread gentle subsidence to the north of the Hercynian mountains during their peneplanation. However the areas of deepest subsidence are underlain by continental rifts such as the precursor of the Viking and Central Grabens (Kent, 1975) and the Oslo Graben and its southern extensions (Illies, 1969). Although volcanic rocks are not common in the Zechstein sequence, volcanism associated with rifting is known from the northern North Sea, the margins of the Fyn-Ringkøbing high and, of course, the Oslo Graben.

The White Pine copper deposits of northern Michigan are contained in a thick volcanic and sedimentary prism that accumulated in the Keewanawan rift (Norman, 1977) in Upper Proterozoic times. A thick essentially mafic volcanic pile first accumulated in the rift and is overlain first by continental clastics (Copper Harbour Conglomerate, Fig. 8.5), then by a marine transgressive series of carbonaceous shales (The Nonesuch shale) and finally by a series of shales and sandstones. There is no chemical unit above the cupriferous Nonesuch shale and, although the basal clastics are red, there is no clear evidence in the sequence of significant evaporitic deposits. Although the bulk of the mineralisation is in the Nonesuch shale the underlying mafic volcanics and clastics are mineralised, particularly beneath richer sections in the shale. The mineralisation clearly cross-cuts bedding and there is much evidence of replacement of earlier pyrite by copper minerals (White & Wright, 1954, 1966; Brown, 1978). The age, setting and style of the Coppermine river area mineralisation (Fig. 8.5) in Arctic Canada is remarkably similar (e.g. Kirkham, 1974). It should be noted that copper is the essential economic metal in these two areas and that lead and zinc are subordinate or even absent. This is in strong contrast to the central African and European examples.

Caia (1976) and Van Eden (1978) have described the numerous Cu (with both lead and zinc in places) mineralisations in Cretaceous strata on both sides of the South Atlantic (Fig. 8.5). The mineralisation is closely associated with the deposits of the first marine transgression over the newly generated and subsiding continental margins. The principal mineralisations are associated with sediments laid down under arid climatic conditions. The underlying clastic strata contain debris of deeply weathered cratonic material and are preserved essentially in half grabens. Their thickness varies radically longitudinally and there was perhaps some 100 m of relief on the basement. The transgressional sediments are limestones, evaporites and carbonaceous shales, which locally contain stratiform copper and iron sulphides. The shales are overlain by, and in part lateral equivalents of, extensive carbonates and evaporites which are in turn overlain by normal marine sediments.

Copper mineralisation is abundant but rarely economic. Although stratiform mineralisation is confined to the carbonaceous units, slightly cross-cutting patches of sulphides and secondary minerals also occur in the basal clastic sequences. Both types of mineralisation are best developed above the thicker basal clastic sequences. In general much of the mineralisation is post-sedimentary although primary sulphides, essentially pyrite, may have formed in some of the euxinic environments.

From these descriptions it would seem that there are really two groups of these deposits: those in which there is a thick basal sequence of mafic volcanics but no necessary indications of aridity during deposition (e.g. White Pine) and those closely associated with arid depositional conditions with no necessary presence of volcanic rocks (e.g. Kupferschiefer). All the other generalisations stated earlier hold for both types. Mineralisation associated with volcanic-bearing sequences is in general less stratiform and occurs in a greater variety of lithologies.

It is also clear that most authors now consider these deposits as diagenetic with mineralisation being precipitated from migrating ground waters preferentially in beds already enriched in syngenetic or earliest diagenetic pyrite, but also occurring in other environments where reducing and sulphurous conditions pertained during diagenesis (e.g. Brown, 1978). We must identify a source for metals and sulphur. Binda (1975) proposed that at least some of the sulphides in Zambia were detrital. The general instability of base metal sulphides under atmospheric conditions lead many workers to doubt a detrital origin. However Clemmey (1978) describes recent copper deposits in the Atacama Desert where clastic copper and iron sulphides from the El Salvador porphyry deposit and mill are mined from fluvial deposits. Detritus reaching the coast is reworked and further concentrated in littoral placers. In these the minerals become oxidised after deposition and both sulphide and oxide-sulphate-carbonate assemblages are mined. Were this desert to be transgressed by the sea these deposits would probably be covered by lagoonal carbonaceous, calcareous and evaporitic strata (e.g. the deposits of the nearby Boca de Vitrila) and then by normal marine sediments (Fig. 8.5). Redistribution of the sulphides and perhaps further concentration would be a further probability during diagenesis in such a pile. A detrital origin then is quite possible but it seems unlikely that the huge tonnages of copper in many of these deposits could have originated in such a way and the presence of large primary copper deposits in the hinterlands of all the various examples is unlikely.

Other workers (Garlick, 1961; Van Eden, 1974) suggest that the copper was supplied in solution from rivers draining the source areas. Garlick further maintains that the copper was precipitated syngenetically, whereas Van Eden subscribes to the diagenetic remobilisation model. Garlick suggests biogenic reduction of sea-water sulphate as the source of precipitant suphur and indeed many of the deposits contain sulphur with isotopic values typical of biogenic reduction (e.g. Burnie et al., 1972). However biogenic reduction of sulphate need not be syngenetic and can occur readily in lithified strata. Annels (1974) notes that in some deposits there is an inverse correlation between mineralisation and unreplaced sulphates in overlying evaporites. He proposed that cupriferous solutions migrated through porous strata into reduced horizons where evaporite sulphate had already been reduced, but begged the question of the ultimate source of copper.

Continental red sandstones would be a good source of copper (adsorbed onto haematite) which could be released to connate water being driven laterally and vertically out of the system during basin evolution and compaction. Migration would be enhanced along palaeochannels and in thicker clastic prisms. The fluid flow would be concentrated at feather edges of the clastic sequences—the ancient shorelines. Such sandstones would be preferentially enriched in copper over lead and zinc, perhaps explaining the preponderance of copper in these deposits. However the enormous tonnages in many of these deposits would seem to require some primary enrichment of copper in the source area. Both the evaporites and early sulphide concentrations in carbonaceous shales provide ideal sulphur sources and sinks for this copper.

The source for copper in the systems with extensive mafic volcanics would seem to be the volcanics themselves; they often contain copper deposits and most workers have accepted them and/or their detritus as source rocks. White (1971) and Brown (1978) have summarised the evidence for diagenetic deposition in these deposits. A source of sulphur is present in both connate water and in the early sulphides in carbonaceous shales. However these volcanic-associated deposits would seem to have fewer and less voluminous sulphur sources than those associated with evaporites. This may explain their higher metal:sulphur ratios. Again the volcanics would be a better source for copper than lead and zinc. The problem of metal source can be further examined using the nickel/cobalt ratio. In general this is greater than one where the metals have been derived through sedimentary processes. In virtually all red-bed coppers the ratio is less than one—a characteristic of metals derived from an igneous source.

In summary a polyphase model for these deposits is widely accepted. Primary sulphide enrichment in euxinic facies; descendant supply of sulphur from connate, marine or evaporite waters and ascendant supply of metals from migrating basinal brines which leached volcanic and/or terrigenous clastic sequences are all involved (Davidson, 1965). There is also the possibility that the brines could have escaped onto the sea floor at times to form syngenetic sulphides in euxinic environments. Mineralisation was usually an early event in the lithification history of the host rocks: it took place at low temperatures and generally predated migration of hydrocarbons, which could not therefore have been a source of sulphur. These conditions are frequently met explaining the widespread coincidence of copper mineralisation and red beds. However certain geological conditions must have enhanced one or more of these genetic processes to make the huge enrichments that are our major mining areas. One of these conditions is for widespread deep weathering in arid conditions of the cratons prior to transgression. Presumably this releases the copper from the rocks but does not remove it from the system. Widespread deep weathering is characteristic of large stable cratonic areas and aridity is typical in the interiors of such cratons. For these deposits with evaporites a restriction to within 30° of the equator (Gordon, 1975) is a further restraint. Large cratonic areas are a consequence, in the Phanerozoic at least, of continental collision. It is not surprising that the greatest abundance of Phanerozoic red-bed copper deposits is Permian, post-dating the aggregation of all the major continental plates (e.g. Smith *et al.*, 1973). Whether or not plate tectonics operated in the Precambrian large cratons certainly existed in the later Proterozoic (Piper, 1976).

The second requirement is for marine transgression with a basal euxinic sequence. Reduced shallow water sequences are again typical of arid and evaporitic environments. Transgression over previously stable cratonic areas requires some sort of rifting. Rifting often follows orogeny (e.g. Central and Western Europe in the Permian) but is also common in the central parts of large continental plates (e.g. Keewenawan Rift) occasionally leading to the formation of new oceans (e.g. South Atlantic). Whether the rifting alone, allowing marine transgression, is enough or whether basal mafic volcanic rocks are needed as a source for copper is more debatable. The generally high cobalt:nickel ratios would indicate the latter. There is therefore a clear association between geotectonic setting, palaeogeography, climate and large red-bed copper deposits.

SANDSTONE URANIUM–VANADIUM–COPPER DEPOSITS

In many texts red-bed copper deposits are often taken to include a class of cupriferous deposits in sandstones in which, uranium and/or vanadium may be extensively enriched. This ore type is however quite distinct in occurring exclusively in arenites, whereas in red-bed copper deposits *sensu stricto* the mineralisation is predominantly in argillites. Uranium-vanadium-copper sandstone deposits are widespread in both time and space, but, although often very richly mineralised, are rarely of large tonnage. However, certain areas contain large numbers of them and form important mining districts. The Colorado Plateau is one such and gives its name as an alternative for these deposits. New Mexico, Utah and Wyoming also contain numerous examples. Elsewhere the oldest so far known occur in the Great Slave Lake area (2.2-1.8 billion years. Morton,

1974; Walker, 1977; Stanworth, 1980; Badham & Stanworth, 1975): the Beaufort Basin in South Africa has recently emerged as an area of some importance (Stapleton, 1978); and both the New (Barthel, 1974) and Old (e.g. Orkney) Red sandstones of Europe contain examples.

The deposits are clearly secondary with the oxides and sulphides precipitated at reduction fronts from oxidised formation waters migrating through permeable hosts. The reductants are often organic debris and include fossil wood, oil shale horizons and even hydrocarbons that had themselves migrated into the host rocks. The deposits are particularly common in fluviatile continental clastic sequences where formation water migration is enhanced in paleo-stream channel deposits. The restriction in space to both channels and reduction fronts accounts for the small size and high grades of these deposits.

The formation waters carrying the metals were oxidised and there is abundant evidence to suggest that in many cases they were meteoric in origin (Fig. 8.4B). They presumably obtained the metals during surface or subsurface oxidative chemical weathering. Although all rocks contain uranium, copper and vanadium, enriched source areas may enhance mineralisation: granites and felsic volcanic strata are relatively enriched in uranium and thorium. It is no accident that granites are common in the source areas and felsic volcanics occur both in the source areas and interbedded with mineralised strata.

The oxidative solution of uranium, its transport and deposition in reduced areas is a very common reaction. In Phanerozoic rocks resultant ores are common in late- to post-orogenic continental arenites derived from and deposited in close proximity to magmatic mountain belts. In the mid-Proterozoic widespread weathering of gneissic cratonic areas both sub-aerially and beneath thin but hugely extensive veneers of sandstones (e.g. the Athabasca sandstone in north Saskatchewan) enhanced these processes and large, rich uranium ores are developed in fissures and regolith in the basement and patchily in the cover (e.g. north Saskatchewan and north Australian examples: Morton, 1976; Hegge & Rowntree, 1978; Dahlkamp, 1978). Thus although there may be superficial similarity between red-bed copper deposits and sandstone uranium-vanadium-copper deposits they are in fact quite different in both genesis and environment of formation.

SEDIMENT HOSTED ZINC–LEAD DEPOSITS

Introduction
Although there is a great variety of stratiform, stratabound zinc-lead deposits in sediments, new discoveries and new ideas seem to be leading

to an understanding that the deposits are all part of a continuum that is related essentially to the migration of metal-charged connate waters (and often hydrocarbons) out of basinal sedimentary prisms during their compaction. The migration may be early or late, fast or slow, voluminous or insignificant. As a consequence mineralising fluids may accumulate in sedimentary, diagenetic or lithified environments and deposition may take place wherever the fluids encounter suitable physico-chemical conditions. The sulphides may therefore occur in sedimentary, diagenetic or epigenetic environments: hence the great variety of deposits. The essential aspect of them all is that they contain predominantly zinc and lead sulphides, usually with fluorite and/or barite and are usually deficient in copper: this we shall have to explain.

Previously many workers (e.g. Stanton, 1972) have recognised two separate classes of zinc-lead deposits: black shale-hosted syn/diagenetic massive sulphides and carbonate-hosted dia/epigenetic massive sulphides. Recent research on the relatively newly discovered Irish zinc-lead deposits has shown their essentially diagenetic nature and has lead to the hypothesis that they form the middle ground of a continuum with the other two types. At one extreme of this continuum early/fast migration of connate waters may cause their escape from the sedimentary prism to surface where they have the potential to precipitate syngenetic sulphides. At the other extreme late/slow migration may cause mineralisation in pore spaces of lithified strata. Although the chemistry of solution, migration and precipitation may be relatively simple the geotectonic conditions which favour one or the other type of deposit may not be. The three main stages in the continuum are dealt with separately and the genetic and geotectonic controls are covered in the summary.

Epigenetic carbonate zinc-lead ores
These ores, often called Mississippi Valley type from their abundance in central USA, characteristically consist of sphalerite and galena with marcasite/pyrite, barite and fluorite in dolomitic carbonates. The mineralisation occurs as open-space filling in vuggy dolomitised or solution-brecciated carbonates. Mineralised carbonates commonly form part of a thick carbonate pile, but are close to facies fronts with basinal argillaceous rocks. There is also a strong spatial connection with evaporitic strata, but no indication of any associated igneous activity. The mineralisation was precipitated from saline, chloride brines at temperatures between 50 and 150°C by sulphur derived mainly from sea water or evaporites. Many deposits contain degraded hydrocarbons and fluid inclusions in minerals often contain hydrocarbons as well. Lead in the deposits is generally of crustal origin being enriched in the radiogenic

isotope: in some deposits there may be addition of lead from primary igneous sources, but this appears to be an accidental enhancement rather than a requirement.

The deposits tend to occur in clusters in favourable districts. These are concentrated on basin margins or around highs. Within them there is often a metal zonation indicating transport from laterally migrating solutions. Various zonation parameters and the close association of oil and ore have lead to a model of lateral (with or without vertical) migration of connate waters and oil out of shale basins during their lithification and compaction along permeable conduits into various stratigraphic and/or structural traps. Oilfield brines are frequently metalliferous and of the same composition that precipitated the ores (Hitchon, 1977; Carpenter *et al.*, 1974). The only real difference between oil and ore in these systems is that the ore requires a precipitant and is then fixed whereas the oil can continue maturing and migrating. Excellent reviews of these ore types and their relationships with oil are by MacQueen (1979), Heyl *et al.* (1974), Jackson & Beales (1967) and a number of important papers in Garrard (1977).

Connate brines contain all the metals found in these deposits but little sulphur. Associated hydrocarbons may contain H_2S or sulphur in mercaptans. However the instability of metal chloride complexes at low temperatures in the presence of reduced sulphur indicates that sulphur was introduced to the brines independently and at the site of deposition. Sulphate in trapped sea water can be reduced biologically or inorganically (e.g. by alkanes). Carbonates themselves are often rich in organic material and H_2S and may therefore contain sulphur awaiting the arrival of metals. Evaporites are rich in sulphates and refluxing brines could supply sulphur to nearby carbonates. There is therefore no shortage of potential independent sulphur sources.

The hydrocarbons and mineralisations in middle Devonian strata of north-west Canada illustrate nicely the relationships between oil and ore and between shale, limestone and evaporite. In fact the facies front between shelf and basin was stable over a relatively narrow area throughout much of Ordovician to Lower Carboniferous time and there are sulphide mineralisations in carbonates of many ages. In the Devonian, however, the shelf behind the carbonate front was the site of deposition of an extensive evaporite sequence covering much of Alberta and Saskatchewan. (Fig. 8.6). It has been proposed (e.g. Hitchon, 1977) that evaporite brines refluxing into the carbonates were primarily responsible for extensive dolomitisation which enhanced porosity and thus permitted greater volumes of fluid flow from the shale basin. It is thus not surprising that dolomitised Devonian carbonates are an important reservoir

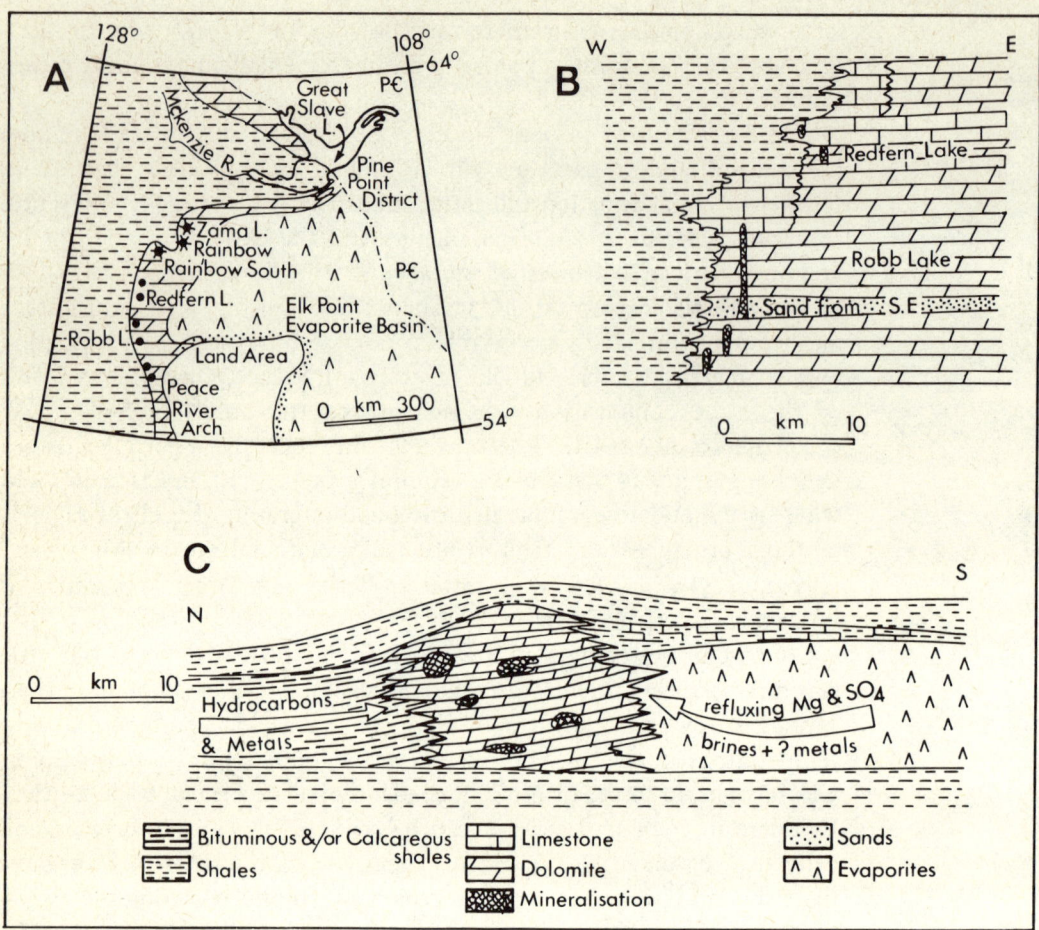

Fig. 8.6. Oil and Pb-Zn mineralisation at facies fronts and close to evaporites in the Devonian of Western Canada. A. The paleogeographic setting with sulphide and oil occurrences marked. B. Schematic section of mineralisation in the Rocky Mountain facies front (after MacQueen & Thompson, 1978). C. Schematic section of the Pine Point mineralised district (modified after Beales & Jackson, 1966).

for hydrocarbons throughout the region and that they host much of the zinc-lead mineralisation of the region. Although mineralisation is common along the entire length of the facies front (Fig. 8.6) it is particularly abundant at the inner end of a recess caused by the MacDonald Fault. The greater volume of ores in the Pine Point district (Fig. 8.6) presumably result from channelled, and therefore more voluminous, fluid flow through the reefal carbonates in this area, and perhaps their greater access to brines from the evaporites—many of the other mineralisations had a more restricted connection to the evaporites because of the intervening Peace River Arch.

The chemical system then is fairly simple and should have operated in a variety of geological environments and throughout time. There are however four main problems.

170

Ore Deposits in Sediments

1 Why are the ores always in carbonates and not in other porous strata connected to shales at facies fronts? In fact there are a few notable exceptions—for example the sandstone hosted zinc-lead deposits of the Caledonide margins in Scandinavia, particularly Laissvall and Vassbo (Christofferson *et al.*, 1979; Ncube *et al.*, 1978; Rickard *et al.*, 1975, 1979). Carbonate sequences have a number of distinct advantages however. First, they are usually lithified early and have a high initial porosity and permeability. Biohermal, shelf-edge systems are particularly good examples of this. These incompactible fluid conduits and reservoirs are closely connected to both fore-reef shales and back-reef (often evaporitic) sequences and can thus easily accommodate the products of their (greater) compaction (MacQueen, 1979). Arenaceous sequences are more compactible, and so lose their primary porosity; they are also usually connected to the shales by sediments with an increasing clay content which militates against good lateral permeability. Second, carbonates are particularly prone to development of secondary porosity and permeability either during dolomitisation or because of weathering (karstification). Dolomitisation is particularly enhanced by refluxing magnesium-rich brines from back-reef evaporites. Development of secondary porosity is not a characteristic of sandstones. Third, fluids migrating into carbonates encounter an environment of totally contrasting pH—a factor which may play an important part in sulphide precipitation. Finally carbonates are often enriched in reduced sulphur which may aid precipitation of metals. Because carbonate sequences usually contain little iron the sulphur is free whereas in sandstones the iron content is sufficient to take out most of the sulphur as pyrite. The suitability of carbonates over sandstones is clear and the exception of the Scandinavian examples lies in their rather sharp facies front onto shales, their low argillite and iron contents and their high organic content.

2 Why are evaporites so commonly associated with carbonate zinc-lead mineralisation? The association is so common that Dunsmore & Shearman (1977) maintain that it is a requirement of successful mineralisation. Certainly evaporites can provide significant volumes of refluxing brines for the carbonates. These may contain magnesium, to aid dolomitisation and thence porosity; SO_4 which may be reduced in the carbonates to provide a precipitant; base metals, particularly in interbedded shale deposits (cf. red-bed deposits), which may provide an alternative to fore-reef shales as source rocks (see e.g. Thiede & Cameron, 1978); and $CaCl_2$ which, even more than NaCl, may enhance the formation of water-soluble metal chloride complexes (see e.g. Hitchon, 1977, p.15). The association of evaporites though not a prerequisite would appear to be beneficial in many ways.

Chapter 8

3 Why is copper essentially missing from these deposits? In part the answer must lie in the source: shales are commonly enriched in most base metals but study of the contents of elements which would be precipitated under relatively cold, reducing, alkaline conditions in limestones shows an order of abundance barium, zinc, fluorine-boron, nickel, strontium-copper, lead. Thus shales are preferential hosts of barium, zinc and fluorine over copper and lead. This explains the general enrichment of zinc over lead but not the almost total absence of copper. A second factor is solubility and temperature. Both empirically and theoretically there is an order of decreasing solubility from copper-zinc-lead in chloride brines. At higher temperatures however the solubility of copper becomes appreciable. Such temperatures ($\sim 200°C$) are not generally reached in these systems—where they were (e.g. Irish deposits, see below) copper does become an important constituent. Thus the whole system in theory, experiment and observation precludes significant copper mineralisation. The significance of source rock is also apparent from study of red-bed copper deposits: presumably they contain little lead and zinc because red-beds and mafic volcanics were better sources of copper. The small amounts of lead and zinc in the Cretaceous of the South Atlantic system already described (Van Eden, 1978) serve only to support this deduction.

4 Why are these deposits abundant in Phanerozoic strata but rare in the Precambrian? Often such temporal variations are more apparent than real—the result of differential preservation, exposure and exploration effort. However, in this case the difference is quite real. There are huge volumes of Proterozoic carbonates in many areas of the world with virtually no associated mineralisation. In Precambrian sequences bioherms and shelf-edge complexes are usually made up of stromatolite rather than coral-dominated assemblages. Most stromatolitic carbonates were lithified very early with no primary porosity. In addition many were primarily dolomite. Thus the essential answer to the problem is that the potential reservoir rocks were impermeable. If this were the case then presumably the fluids had to migrate somewhere: they could have precipitated their metals in shales adjacent to the carbonates, or could have been forced upwards to surface, there to precipitate metals in suitable environments. Shale-hosted zinc-lead mineralisation is not uncommon in Precambrian strata (see below).

The model for the origin of these epigenetic deposits is well supported and generally accepted. We must however assess which factors may enhance mineralisation. Time would appear to be one, with the development of coral-dominated shelf carbonates being important. Shale-carbonate facies fronts are a requirement, but are abundant both at continental margins and in epicontinental marine sequences. The presence of evaporites close to the carbonates appears to be a significant factor in enhancing ore deposition: the evaporites may play a role in host rock preparation, in sulphur supply and possibly even in metal supply. Thus climate is an important factor—Van de Poll (1978) noted the importance of arid intervals in the mineralisation of Carboniferous shelf strata in east Canada. The relationships between oil and climate were made clear earlier in this book (Chapters 2-4). So too were the geotectonic conditions that enhanced oil formation and accumulation—and those same geotectonic conditions apply to the mineralisations (Chapters 6-8). Within the framework of shelf-basin facies fronts, oil and ore will be enhanced wherever the basinal sediments underwent a suitable maturation history. Strata that never got warm enough will never have released oil or mineralising fluids. In strata that got too hot only thermal gas or graphite will form but mineralising fluids could still be generated. For hydrocarbon accumulation the reservoir areas too must remain below about 200°C, whereas sulphides, once precipitated, are fixed and will undergo metamorphism along with their hosts.

To get large oil and/or ore accumulations the prime enhancement factor is to concentrate fluids from a large source volume into a small reservoir volume. Consequently thick basinal sequences in fault-bound depressions, especially at continental margins, will have their fugitive fluids channelled along the line of the faults. Such is clearly the case at Pine Point. Continental margin fault-bound recesses (aulacogens) are commonly magmatically active as well. While the extra magmatic heat may 'over-mature' the rocks as far as hydrocarbons are concerned, it will enhance fluid circulation and leaching and thus encourage mineralisation. Sawkins (1976) argues that many Mississippi Valley-type deposits are related to hot spots and cratonic rifting, citing particularly the small deposits of the Benue Trough (Olade, 1976). Extensions of his argument to central USA are less well supported and more tendentious especially in his hint (p. 661) that peralkaline magmas were needed as a source for lead, zinc, barium and fluorine. Sharp (1978), considering the Ouachita Basin and Ozark mineralisation, supports the graben model. He proposes that pulsatory tectonic de-watering of thick Palaeozoic flysch deposits caused the lead-zinc-fluorine-barium mineralisation but saw no need to invoke magmatic rocks either as a metal source or as a hydrothermal engine. Brecke (1979) postulated a similar process for the same mineralisation but presumed it to be Late Mesozoic and used Cretaceous igneous rocks as the dominant driving force. In general there is so rarely any connection between these ores and magmatic activity that one can hardly accept that, where there is a coincidence of the two, it has any genetic significance. The details of these various hypotheses may differ but their

Chapter 8

common aspect is clear, that extra thick sedimentary piles generate more and better ore fluids and that where these are channelled by faults during lateral migration large deposits may result. These conditions are met particularly on rifted unstable cratons and at faulted indentations in continental margins.

Irish-type zinc-lead deposits

In the preceding section lateral migration of normal formation waters was identified as the prime cause of carbonate-hosted deposits. Throughout there were hints that higher than normal thermal gradients would cause a number of important changes to the system. First, fluids would be driven further laterally and possibly vertically as well: thus vertical conduits could become important. Second, the relationship between oil and ore would be lost with the thermal destruction of hydrocarbons. Third, the leaching capacity of the formation waters would be enhanced such that they would also start to leach and carry significant amounts of copper.

Major zinc-lead deposits occur in Lower Carboniferous carbonates in a number of places in Ireland. They include Tynagh, Silvermines and Navan, the latter containing *at least* 70 million tonnes of 13% zinc + lead. All these deposits and many of the other mineralisations (Fig. 8.7) are in carbonates of the Waulsortian facies which consist of biohermal carbonate mudbanks deposited in a warm, shallow sea between land and fringing areas, generally to the north and east, and deeper water carbonaceous shales to the south and west (Fig. 8.7). The near shore deposits contain evidence of evaporites and algal tidal flat deposition (West *et al.*, 1968). During the Lower Carboniferous many of the major Caledonian faults were reactivated and mark lines of important facies and thickness change. All the deposits lie on or close to these fault lines. The deposits also preferentially overlie areas of thicker Palaeozoic sediments—both the conformably underlying Old Red Sandstone facies and the earlier metamorphosed Caledonian strata. None of the deposits shows any spatial connection with volcanism even though alkaline volcanic activity was taking place, especially in the Limerick area.

The deposits are mineralogically and paragenetically complex. In addition to the usual zinc-lead-barium-fluorine they contain extensive copper, arsenic and antimony and the odd deposit at Gortdrum was mined for copper-mercury-silver alone. The deposits are also often much more pyritic. In all there is clear evidence of very early sulphide and sulphate deposition (mostly pyrite, sphalerite and barite) on the sea floor (e.g. the Silvermines Upper G zone—Taylor & Andrew, 1978), in unconsolidated organic-rich lime muds and in fault-controlled fractures.

Ore Deposits in Sediments

Fig. 8.7. Palaeogeography of the Lower Carboniferous of the British Isles, showing the location of Waulsortian facies and associated mineralisation in Eire and the association of mineralisation with block margins in England (modified from Boast, 1979; and from Moore, pers. comm.).

In some places, although deposition of sulphides took place within the sediments, the spent fluids precipitated oxides (haematite and chert) on the sea floor (e.g. the iron formation and extensive manganese aureole at Tynagh-Russell, 1975). The faults appear to have acted as conduits for this early mineralisation. Later faulting caused extensive brecciation in many of the deposits and was coincident with the introduction of more copper- and lead-rich fluids.

Although the early mineralisation may have taken place at relatively low temperatures there is clear evidence from mineral textures and assemblages, from isotopes, from fluid inclusions and from the lack of association with liquid hydrocarbons that depositional temperatures were mostly in the range 200-350°C (Boast, 1978; Morrissey, 1977; Coleman, 1977).

Russell (1978) and Radtke & Russell (1978) have suggested that the ore fluids were connate waters released from underlying lower Palaeozoic clastic sediments and from the Old Red Sandstone: in effect they say that the fluids are more than just the products of compaction de-watering and are essentially fugitive metamorphic waters. The enhanced copper content could be explained by the higher fluid temperatures but the increases in mercury, arsenic and antimony may well indicate that volatiles from alkaline magmas had a part to play too. Boast (1979) clearly suggests a magmatic origin for late-stage lead in the deposits.

It would seem that the ore fluids were generally similar to those causing Mississippi Valley-type deposits but that they consistently reached the diagenetic and occasionally the syngenetic environment. The paragenesis of Irish deposits appears to span the entire depositional and diagenetic history of the Lower Carboniferous strata. Precipitants for the metals were mainly biogenetically reduced sea-water sulphate in the limestones and Boast (1978) suggests that early mineralisation was fixed by sulphide generated *in situ*, while later stages required lateral migration of reduced sulphur through the Waulsortian limestones. However, the final stages in the paragenesis, especially those deposited nearest the faults, have magmatic S-isotope values, again indicating the late influx of deeply derived, magmatically influenced fluids. There is no indication that evaporite brines were of any significance.

There are two critical factors differentiating these Irish deposits from the purely epigenetic ones. First, the faults that acted as conduits penetrated deeply enough to tap a variety of formation, magmatic and metamorphic fluids. Second, the high heat-flow and tectonic instability of the area enhanced leaching properties of fluids, migration rates and migration distances—vertical and lateral. In fact the whole area from Nova Scotia to western Europe was a region of block-faulting, high heat-flow and local rapid subsidence and accumulation. It is also characterised by numerous lead-zinc mineralisations—Gays River district in Nova Scotia, the Irish province, the Mendips and Pennines and deposits throughout Belgium and southern Germany. While many of these are epigenetic all show clear evidence of the upward movement of fluids along growth-faults particularly at basin margins. In general the fluids did not quite have the energy to reach the sea floor and many of those that did probably encountered oxidising conditions and were dissipated.

The geotectonic setting for the Irish deposits is less clear. Russell (1976) proposed that incipient oceanic rifting of the newly formed Eur-American plate was responsible for the instability, local magmatism and high heat-flow. In general this seems an unlikely setting in an area that had just been involved in major collision orogeny and which was still suffering the continued stress of the arrival of the African plate. Taphrogenic relaxation with associated volcanism and block-faulting is a more plausible setting: this, enhanced by the jostling of cratonic blocks induced by the dextral shear couple responsible for the Hercynian orogen (Badham & Halls, 1975; Arthaud & Matte, 1976) to the south, could well have driven formation waters up fractures to the surface.

In these Irish deposits the tectonic controls become dominant and the climatic ones are only important in so far as Waulsortian facies deposition (generally tropical) was a requirement.

Shale-hosted zinc-lead deposits

If, under exceptional circumstances, formation waters can be driven out of lithifying sedimentary piles and into the diagenetic environment and even onto the sea floor there is no reason why carbonates are a required associate. Sea water itself is alkaline enough to induce instability of the metal complexes and all that is needed is a supply of reduced sulphur; and reduced sulphur, as we have seen already, is abundant in carbonaceous shales, particularly those associated with deep, euxinic basins or with shallow hypersaline waters.

Black shales host some of the world's greatest ore deposits notably: Sullivan (>170 mt. 9% Pb+Zn); McArthur River (>190 mt. 13.6% Pb+Zn); Howard's Pass (~400 mt. 10% Pb+Zn); Mount Isa (~230 mt. 13% Pb+Zn). Smaller but historically important deposits include those of Meggen (Ba-Zn), Sardinia (Pb-Zn-Cu-As), the Pyrenees (Zn-Pb), northern Spain (Zn-Pb) and Sargipali, India (Zn-Pb-Cu-As). They can be divided essentially into those associated with deep-water sediments (Sullivan, Howard's Pass, Meggen, Pyrenees) and those in relatively shallow-water sediments with associated carbonates, evaporites and chert (Mount Isa, McArthur River). These deposits are by no means common, but are more abundant in Proterozoic strata. However, Archaean environments were unsuited to mineralisation of this type, and the disproportion between Proterozoic and Phanerozoic examples is more apparent than real as the Proterozoic lasted over three times as long. Despite their immense size, their common preservation without any major deformation or metamorphism and their clear stratabound form, there is perhaps more controversy over these deposits than any other. Both syngenetic exhalative and epigenetic replacement models have been proposed and the syngeneticists are divided into those who require volcanic impetus for exhaled fluids and those who do not. Some of these deposits and their settings will be described before examining the genesis controversy.

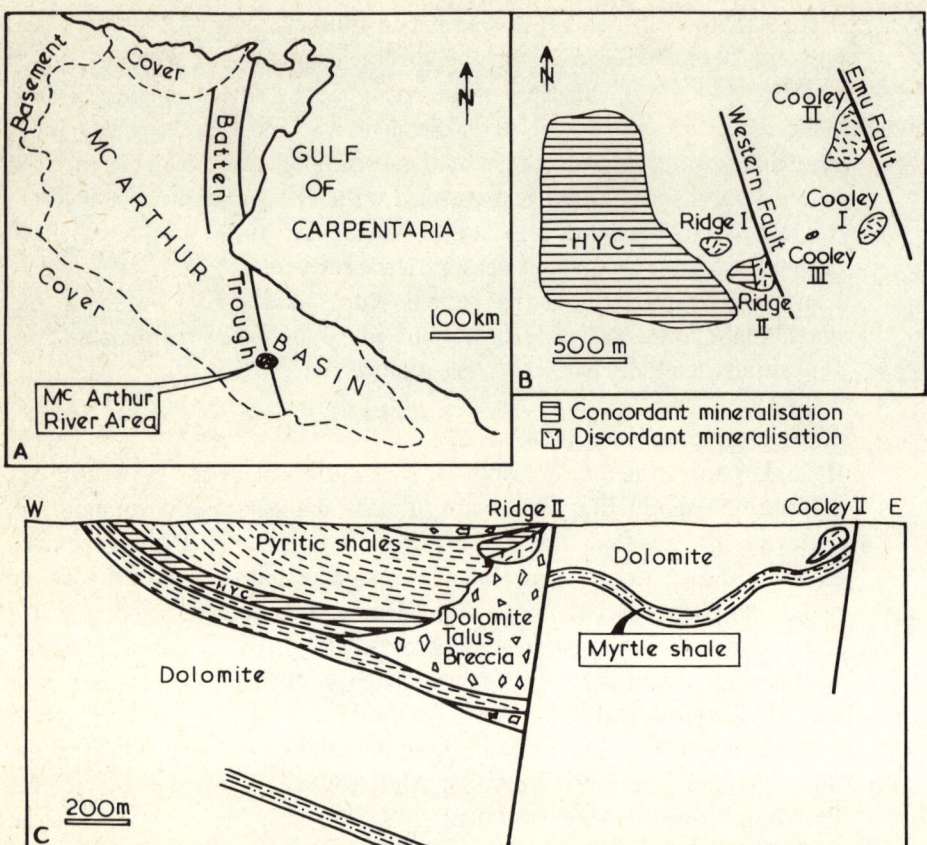

Fig. 8.8. Geology of the McArthur River mineralisation. A. Regional map showing the relationship between mineralisation and the Batten Trough. B. Local map showing the locations of faults, concordant and discordant mineralisation. C. Cross-section showing the nature of the two types of mineralisation and their relationships to the faults (modified from Williams, 1978).

Of shallow-water shale-hosted deposits those of the McArthur River area are so little deformed and well preserved that they must be described as the type example. The deposits lie in a Middle Proterozoic (∼1600 million years) sequence of dolomites, relict evaporites and carbonaceous shales (Cotton, 1965; Croxford *et al.*, 1972, 1973, 1975). These sediments were deposited in the widespread McArthur Basin but more particularly on the edge of a fault-bound trough marking the deepest parts of this basin (Fig. 8.8). The deposits include the immense stratiform HYC pyritic lead-zinc deposit and a number of slightly discordant copper-lead-zinc lenses particularly near the faults. There is much evidence that the faults were active during sediment and sulphide deposition. The controlled thickness and facies changes and also clearly acted as conduits for ascending mineralising fluids (Williams, 1978).

The stratiform ores are all contained within pyritic, carbonaceous shales and are characterised by delicately banded sulphide layers with intervening shale. Successive sulphide and shale layers are chemically and mineralogically distinct and Croxford *et al.* (1972) argue that they could only have originated by direct primary sedimentation. Williams (1978) presents extensive evidence that pyrite in the shales was present as syngenetic/earliest diagenetic layers but that it was overgrown later by base metal sulphides deposited from solutions migrating along bedding. He supports his case by showing that deposits in and near the faults are discordant and richer in copper, and that these give way laterally to concordant deposits with more lead and zinc. It should be noted that all the data at present come from drilling so that actual continuity from concordant to discordant cannot be properly documented. From the data available the HYC deposits are best regarded as syn/diagenetic and the discordant deposits as epigenetic. Overgrowth of pyrite by base metals during earliest diagenesis is a common feature of many syngenetic deposits and is not a requirement of epigenesis. The extraordinarily delicate but varied mineral banding is hard to explain by any replacement model and there is no obvious reason why, if solutions emanating from the faults were able to penetrate shales, they did not penetrate and mineralise the much more porous surrounding dolomites. That said, Williams' contention that mineralising fluids did emanate from the faults and were responsible for the discordant mineralisation is equally valid. There is no reason why they should not have emerged at surface and spread laterally into the shallow, hypersaline, reducing shale basin to precipitate their metals. The fact that some discordant deposits are stratigraphically above concordant ones poses no problems—one merely has to postulate a long-continued fault and mineralisation history, and there is plenty of evidence of both.

The Mount Isa mineralisation has been severely deformed, but again delicately laminated sulphides are interbedded with sediments and again these sediments are shallow-water carbonates and carbonaceous shales. Here too there is metal zonation with pyrite in the black shales changing laterally into sphalerite and galena, and with only copper and iron sulphides in the carbonates. The carbonates and shales interdigitate and have been interpreted as algal reef mounds and intervening off-reef sediments (Stanton, 1972). The ore body is found in a thick sequence of sediments in a graben and again there is evidence that faults were active during deposition.

The close association of these two ore bodies with shallow-water hypersaline and carbonaceous sediments suggests a close relationship with the red-bed copper deposits discussed earlier. In both examples, however, not only is there clear evidence of rifting and thick sedimentation but there is also evidence of initial bimodal volcanism in the form

Chapter 8

of mafic dykes and numerous felsic shards and tuffs (e.g. Sawkins 1976; Croxford, 1972). Formation waters were driven out to the surface up growth-faults from thick sedimentary prisms by the extra impulsion of rapid sedimentation and magmatic activity. The dominance of lead and zinc over copper is presumably a result of greater leaching of argillaceous sediments and lesser of red beds or mafic volcanics in these particular systems. Otherwise these deposits must be considered to be closely related to red-bed mineralisation.

Deep-water shale-hosted deposits are rather different. The Sullivan and related deposits occur in pyritic shales and siltstones in the lowest formation of the clastic middle Proterozoic Purcell sequence. Preceding and subsequent clastic sedimentation on the Beltian prodelta was relatively constant and well ordered but the ore lenses lie just above a horizon of slumps and conglomerates (Freeze, 1966). The ores are actually confined to areas of restricted sedimentation. There is evidence of growth-faulting and instability; high heat-flow in the abundant Moyie diabase sills of similar age; and for exhalation not only of lead-zinc-iron-rich fluids (Campbell & Ethier, 1977) but also of boron-rich fluids (Ethier & Campbell, 1977). Ethier & Campbell propose that the fluids obtained their metals and boron by leaching the underlying clastic prism and were released up 'pre-ore collapse features', i.e. faults. There is no known direct spatial association with volcanism. The mineralisation is mainly of iron, lead, zinc sulphides with minor copper, arsenic, antimony, tin and silver. The mineralisation is zoned and delicately banded.

The geotectonic setting of the Sullivan deposit is reasonably understood. The mineralisation clearly took place on an Atlantic-type margin. Such margins subside due to cooling following rifting; subsequent blanketing by sediments causes the normal or low heat-flows commonly observed. These areas fail to meet the conditions postulated for exhalation of formation waters. However, Kanasewich (1968) demonstrated the existence of a Precambrian rift subsurface in southern Canada within the extension of which are the Sullivan and Coeur d'Alene mineralisations: he proposed a genetic connection. Harrison (1972) suggested that the Beltian deltaic sequence accumulated in a continent margin rift which Maxwell (1974) suggests was generated during his postulated Mid Proterozoic rifting of western North America. Burke & Dewey (1973) proposed that marginal rifts are common as failed arms of successful spreading systems: such failed arms (aulacogens, e.g. the Benue Trough) may remain volcanically active for some time after rifting (viz. the Camaroun volcanics) and heat-flow in the thick sedimentary prisms may be elevated. Consequently Sawkins (1976) identified aulacogens as suitable environments for base-metal mineralisation and cited the Sullivan as an example. In general aulacogens are sites of rapid clastic sedimentation which may preclude development of large exhalative ores. Continental rifts where initial rapid sedimentation is often followed by prolonged restricted shallow-water deposition are perhaps better suited—cf. McArthur River. Whatever the state of the argument for and against aulacogens a general setting of an unstable, magmatically active Atlantic-type margin seems most apt for the Sullivan deposit.

The Howard's Pass deposit is contained within Ordovician carbonaceous shales on the Yukon-Northwest Territories border. These were deposited in a basin between a carbonate-reef complex to the east and an active volcanic terraine (with related volcanogenic deposits) in the Anvil area to the west (Monger *et al.*, 1972). Underlying Cambrian strata are mostly shallow-water carbonates and a coarsening upwards Silurian-Devonian flysch sequence spread from the west to cover them. The Ordovician represents a period of foundering of the continental margin and of very restricted clastic input—it has a maximum thickness of 120 m in the area (Badham, 1973c).

The mineralisation occurs in thin bands of sphalerite and galena in cyclically deposited highly graphitic mudstones (Morganti, 1977) which also contain thin beds of pyrite, gypsum, barite, chert and carbonates. Copper, arsenic, antimony and silver are present in trace quantities. In all samples measured, the nickel/cobalt ratio exceeds unity. The deposit must have accumulated over a considerable period of time. There is no association with magmatic rocks apart from a few thin tuffs in the Cambrian—presumably wind blown from the arc to the west. However, there is evidence of growth-faulting—particularly at the foundered basin margin.

Badham (1973c, 1975) and Morganti (1977) propose that warm formation waters escaped up faults in and onto an anoxic sea floor and were trapped over a wide area in small, connected depressions where they precipitated their solutes. Other parts of the same belt, southward into British Columbia and northward to Alaska, have great potential for similar deposits. The key factors for mineral deposition appear to have been very low deposition rates and anoxic bottom conditions in a small ocean basin. The delicate thin banding of sulphides and carbonaceous shales over wide areas testifies to the slow and prolonged mineralising process.

The Howard's Pass deposit and others in the Lower Palaeozoic shales of the Yukon and Alaska were probably formed on the continental side of a back-arc basin. The evolution of back-arc basins involves rifting, foundering and rapid infill particularly from the arc side. As the basin widens, so central areas (still with high heat-flow—McKenzie & Sclater, 1968) become distal or separated from sources of clastic sediment. Early erosion on the continental side is likely to cause rapid peneplanation and, in the absence of any orogenic process, the clastic supply from there will also dry up. It is proposed that hot formation waters leach metals in the clastic prisms and migrate towards the continental margin. The higher than normal temperatures of these fluids gives them energy enough to rise up the basin-margin faults and exhale onto the sea floor at a time when sedimentation rates have fallen drastically. Hydrocarbons may also be exhaled with these fluids (cf. Schlanger & Combs, 1975) but the higher temperatures imply an earlier formation and migration of oil and only thermal gases might remain. Continued exhalation onto a relatively stable substrate could give rise to well-banded widespread deposits. Such then is the model which is derived principally for the Howard's Pass mineralisation.

The Meggen barium-zinc deposits lie in a Middle Devonian basinal sequence of shales and micritic limestones. Abundant carbonaceous matter and pyritic nodules in the shales testify to anoxic bottom conditions. The finely banded stratiform nature of the ore and the associated widespread stratiform manganese halo have been used as evidence for a syngenetic exhalative origin (Gwosdz & Krebs, 1977). These authors also identify alteration around conduits beneath ore. There is no direct association with volcanism but high heat-flow in the basin is evidenced by common spilite and/or keratophyre magmatism especially during the Lower Devonian.

The geotectonic setting of the deposit in the Hercynian Orogen is complex. Badham & Halls (1975) and Arthaud & Matte (1976) identify the orogen as the product of dextral shear between major plates. Badham (1976) maintains that such a tectonic system leads to both compression and extension in different areas at different times, the zones of extension being characterised by basin subsidence in block faults, thick sedimentary fill and bimodal magmatism (cf. Crowell, 1974). Munha (1979) proposes similar conditions for the Iberian Pyrite belt, where there is a direct association of volcanism and mineralisation. Burke & Sawkins (1978) also postulate a connection between rifting and mineralisation in the Hercynides, but perhaps with a greater continuity in rifting in both time and space than is apparent in the field.

The Middle Silurian deposits of Sardinia (Zuffardi, 1967) differ essentially only in being very much smaller. There are numerous small syngenetic lenses of iron-lead-zinc-copper-arsenic sulphides associated with carbonate and chert layers in a sequence of highly graphitic shales. Different stratigraphic levels are characterised by slightly different mineralogies emphasising a syngenetic origin. The ores are finely banded and show numerous sedimentary structures. There is no spatial or temporal connection with any magmatic rock. Deposits are concentrated at the margins of a basin of restricted sedimentation adjoining shallow marine and continental sediments. They are particularly concentrated where the Middle Silurian is thinnest and the underlying basement is 'high'. These features suggest that sedimentation was controlled by growth-faulting on the basin margin and that the ores are related to some fluid migration process. Although Zuffardi proves the syngenesis of the deposits he postulated no mechanism—other than a 'purely sedimentary process' (1967, p. 232) for their formation. Exhalation of warm formation waters onto an anoxic basin floor would provide an ideal mechanism. However, the geotectonic setting is poorly understood.

Other deposits which may be similar include the Upper Ordovician zinc-lead-iron sulphide ores of the Pyrenees, the Cretaceous Reocin deposits of northern Spain and the Indian deposits at Sargipali. Some of the Ordovician ores in the Pyrenees do show a spatial and temporal connection with a bimodal volcanic suite (e.g. Pierrefitte; Bois & Pouit, 1976) but others show no such connection and lie conformably in carbonaceous and calcareous phyllites. A syngenetic origin for the deposits from exhaled volcanic solutions is proposed by Bois & Pouit, but they point out that the volcanic connection is frequently absent and exhalation of hot formation waters is certainly a possible mechanism for ore formation. Monseur (1967) argues a syngenetic origin for the lead-zinc-iron sulphide ores in the Reocin area but suggests no mechanism for their formation. There is certainly a possible temporal and spatial connection with the Albian-Aptian rifting of Biscay. The Sargipali deposit (Sarkar, 1974) is in fine-grained calcareous and clastic Proterozoic strata and contains iron-zinc-lead-copper-arsenic with trace silver. Sarkar notes confinement of the ore horizon to a horizon of stagnant, restricted deposition and points out the marked absence of any possible volcanic source.

Many of these deposits which have been described as of sedimentary-exhalative origin have been described by others as volcanogenic. Many workers have gone to extreme lengths to highlight the almost insignificant presence of volcanic rocks associated with some of these deposits and have thus ignored fundamental differences between the two types—the huge lateral extent, most delicate and variable banding and the evidence of accumulation over a long period of time of these shale-hosted deposits. In volcanic terrains fluid expulsion is usually a hot, energetic process and is consequently short lived (Solomon & Walshe, 1979). Volcanic sea floors are unstable and have high relief. Deposits tend to be thick, proximal (in the sense of distance from source—see Large, 1979) and relatively poorly banded. Negative Eh conditions are not prevalent but are created locally by fluid exhalation, so the boundaries (lateral and vertical) between sulphide and oxide-sulphate-carbonate facies are abrupt. Probably the greater tonnage of exhaled base metals is never precipitated but is dispersed because of exhalation into oxidising environments (Badham, 1979).

Exhalation of formation waters into anoxic foundered marginal or rift basins might be expected to last a considerable period of time and to be a relatively docile process. Basin floors are usually of low relief so fluids would spread out over wide areas. Pulsatory exhalation and fluctuations in Eh could cause widespread thin alternating bands of sulphides, carbonaceous sediment and chert, carbonate or sulphate. Sulphur would be predominantly derived by reduction of sea-water sulphate. These features are typical of both types of shale-hosted deposits described above and are quite contrasted to those of volcanogenic deposits.

Conclusion on zinc-lead deposits

Formation waters associated with hydrocarbons are potential ore fluids for iron-lead-zinc (barium-copper). These waters derived their solutes by leaching host strata during burial as they become warmer, more acidic and more saline. Normally such fluids are either dispersed or form carbonate-hosted secondary lead-zinc deposits. In areas of higher than normal heat-flow such fluids may be expelled from the sedimentary column, particularly up faults. If this occurs beneath the sea into a reducing environment sulphides will be precipitated. If expulsion continues for some time and if the sedimentation rate is low considerable volumes of sulphides may accumulate. Such a mechanism is proposed for the formation of both Irish-type and shale-hosted lead-zinc deposits. Unstable subsiding cratonic areas, the continental-margin side of back-arc basins and rifted or volcanically active Atlantean margins are suggested as suitable areas for exhalation of formation waters.

8.4 Conclusions

To conclude this brief review we must ask whether or not there is any relationship between mineralisation in sediments and geotectonics. In the literature one can identify two extremes of opinion. The one (Sawkins, 1976; Burke & Sawkins, 1978) demands a close genetic association between many of the deposits described and rifting—and thence hot spots. On the other hand Sangster (1979) considers many of these deposits to have required the absence of geotectonic processes for their formation—although such a view is rather naïve in that orogenic quiescence is as much a part of geotectonics as is activity.

As was stated at the outset most mineralisations in sediments are themselves simply the result of normal chemical and physical processes that occur during and after sedimentation. The ores may themselves be sediments or they may be the products of diagenesis and lithification of sediments. In so far as sediments and what happens to them after deposition are related to geotectonics then so too are the mineralisations. Such a statement is however of little value either to the student of economic geology or to the exploration geologist. What we must ask is what factors operated during those normal processes to make them abnormal to produce enrichments in metals sufficient for exploitation?

For the physical primary ores processes of erosion, deposition and preservation are the most important. While these may result in ores at any time and any place it is quite clear that the geotectonics of lowermost Proterozoic times favoured the formation of giant placer deposits. Neither before nor since has the combination of enriched source areas feeding a restricted repository with so high a preservation potential been repeated. Similar generalities hold for the Proterozoic Iron Formations. It is impossible to be more specific.

Other controls that have been important in formation of primary sedimentary ores are climate and the natures of hydrosphere, biosphere and atmosphere. It would seem obvious at first that these factors would be of far less importance in formation of secondary mineralisation. All the examples described are the results of migration of water through sediments. Factors enhancing mineralisation are principally richness of source rocks, rates of fluid migration and constriction of migration paths and precipitant areas. For lead and zinc many rocks may be good sources, but shales are particularly so: however, shales are so widespread that there can be no geotectonic control here. For copper however, red beds and basalts are good sources—and both may be especially abundant and available to leaching waters in rift zones. Enhanced migration rates are favoured by extra-thick sedimentary prisms or by magmatism. Again both are common in rifts, particularly those intersecting continental margins. It is faults, and thus rifts, again which may act to channel migrant fluids into restricted areas. In fact the relationship between secondary sedimentary ores and some sort of growth-faulting is so common as to be almost a requirement.

In all the descriptions of secondary sulphide mineralisations the key factor was the supply of reduced sulphur to the precipitant areas—and again and again evaporites seem to have been important in this context. To return the discussion full circle evaporites, by providing high salinity brines, may also enhance leaching rates and metal solubility in the source rocks. So evaporites, and thus climate, are the second important control. But what of the controls on evaporite deposition? Climate certainly is of fundamental importance but there is, too, a close association between evaporites and rifts.

To return finally to the contrasting views of Sawkins and Sangster outlined above—which should we accept? Sawkins is clearly correct in identifying the significance of rifts, but overstates his case by requiring the rifts to be necessarily active during mineralisation. By their very existence in the sub-surface they control sedimentation and fluid migration in overlying sediments during the deposition of which there may indeed be an absence of geotectonic activity. The conclusion must be that there is always a relationship between geotectonics and mineralisation but that this may be so broad as to afford little satisfaction to either academic or exploration geologist.

Acknowledgements

I am grateful to all the mining companies who have over the years allowed me to visit and work on their properties and most generously made available a great deal of information which, although not confidential, might not normally have been published. These companies include particularly Rio Algom Ltd, Vestor Explorations Ltd, Tara Mines Ltd, Irish Base Metals, Consolidated Mogul Ltd, Anglo American Corpn, Kintla Ltd, Bula Ltd, Canex Placer, Cominco Ltd, Canadian Superior and Phelps Dodge Ltd. I am grateful to Harry Clemmey, Bob Hodder, Bruce Bouley, Bill Stanworth, Rick Walker and Roger Morton for discussions on various topics covered in this chapter. I am also grateful for the stimulating discussions with undergraduates who had to put up with the labour pains while taking the economic geology option at Southampton. Anthea Dunkley drew the diagrams and Lola Emery typed and proof read the manuscript.

References

Alexandrov, E.A. (1973) The Pre-Cambrian banded iron formations of the Soviet Union. *Econ. Geol.*, **68,** 1035-1062.

Annels, A.E. (1974) Some aspects of the stratiform ore deposits of the Zambian Copperbelt and their genetic significance. In *Gisements stratiformes et Provinces Cupriferes* (Ed. P. Bartholomé) Soc. Géol. Belgé. Liege., pp. 235-254.

Arthaud, F. & Matte, P. (1977) Late Paleozoic strike-slip faulting in southern Europe and northern Africa: result of a right-lateral sheer zone between the Appalachians and the Urals. *Bull. Geol. Soc. Amer.*, **88,** 1305-1320.

Badham, J.P.N. (1973a) Volcanogenesis, orogenesis and metallogenesis, Great Bear Lake, N.W.T. Ph.D. Dissertation, University of Alberta, 363 pp.

Badham, J.P.N. (1973b) Calc-alkaline volcanism and plutonism from the Great Bear Batholith, N.W.T. *Can. J. Earth Sci.*, **10,** 1319-1328.

Badham, J.P.N. (1973) *Report on the geology and mineralisation, Summit Lake Area, Yukon Territory.* Private report to Vestor Explorations Ltd on open file D.I.A.N.D. Whitehorse.

Badham, J.P.N. (1975) *A new Pb-Zn deposit in the Yukon.* Abstract to Minl-Dep. Studies Group, Leicester.

Badham, J.P.N. (1976) Cornubian tectonics—lateral thinking. *Proc. Ussher Soc.,* **3,** 448-52.

Badham, J.P.N. (1979) Sulphide Ore Deposits. *Nature,* **277,** 513-4.

Badham, J.P.N. & Halls, C. (1975) Microplate tectonics, oblique collisions and the evolution of the Hercynian orogenic systems. *Geology,* **3,** 373-376.

Badham, J.P.N. & Stanworth, C.W. (1977) Evaporites from the Lower Proterozoic of the East Arm, Great Slave Lake. *Nature,* **268,** 516-8.

Barthel, R.H. (1974) Review of uranium occurrences in Permian sediments in Europe, with special reference to uranium mineralisations in Permian sandstones. *IAEA Symposium SM 183-34.* Vienna, pp. 277-288.

Bayley, R.W. & James, H.L. (1973) Precambrian iron formations of the United States. *Econ. Geol.,* **68,** 934-959.

Beales, F.W. & Jackson, S.A. (1966) Precipitation of lead-zinc ores in carbonate reservoirs as illustrated by the Pine Point ore field, Canada. *Trans. Inst. Min. Metall.,* **75,** B 278-285.

Binda, P.L. (1975) Detrital Bornite grains in the Late Precambrian B greywacke of Mufulira, Zambia. *Min. Dep.,* **10,** 101-108.

Boast, A.M. (1978) A textural and isotopic study of Irish base metal mineralisation of Lower Carboniferous age, with specific reference to the Tynagh Deposit. Ph.D. Thesis, Imperial College, London, 208 pp.

Bois, J.P. & Pouit, G. (1976) Les minéralisations de Zn(Pb) de l'anticlinorium de Pierrefitte: un example de gisements hydrothermaux et sédimentaire associés au volcanisme dans le Paleozoique des Pyrénées centrales. *Bull. B.R.G.M. 11,* **6,** 543-567.

Borchert, H. (1960) Genesis of marine sedimentary iron ores. *Inst. Min. & Metall. Bull.,* **640,** 261-279.

Brecke, E.A. (1979) A hydrothermal system in the Midcontinent Region. *Econ. Geol.,* **74,** 1327-35.

Brock, B.B. (1961) Structural setting of the copperbelt. In *The Geology of the North Rhodesian Copperbelt* (Ed. F. Mendelsohn) MacDonald, 523 pp.

Brown, A.C. (1978) Stratiform copper deposits—evidence for their post-sedimentary origin. *Minl. Sci. Eng.,* **10,** 172-181.

Burke, K. & Dewey, J.F. (1973) Plume-generated triple junctions: key indicators in applying plate-tectonics to old rocks. *J. Geol.,* **81,** 406.

Burke, K. & Sawkins, F.J. (1978) Were the Rammelsberg, Meggen, Rio Tinto and related ore deposits formed in a Devonian rifting event? *Econ. Geol.,* **73,** (Abstr.) p. 308.

Burnie, S.W., Schwarcz, H.P. & Crockett, J.H. (1972) A Sulfur Isotopic study of the White Pine Mine, Michigan. *Econ. Geol.,* **67,** 895-914.

Button, A. (1976a) Transvaal and Hamersley basins—review of basin development and mineral deposits. *Minls. Sci. Eng.,* **8,** 262-92.

Button, A. (1976b) Iron Formation as end-member in carbonate sedimentary cycles in the Transvaal Supergroup, South Africa. *Econ. Geol.,* **71,** 193-201.

Caia, J. (1976) Paleogeographical and sedimentological controls of Cu, Pb and Zn mineralisations in the Lower Cretaceous sandstones of Africa. *Econ. Geol.,* **71,** 409-22.

Campbell, F.A. & Ethier, V.G. (1977) Environment of ore deposition for the Sullivan deposits. *Geol. Assoc. Can. Proc.,* Abs. p. 10.

Carpenter, A.B., Trout, M.L. & Pickett, E.E. (1974) Preliminary Report on the origin and chemical evolution of lead and zinc rich oil field brines in central Mississippi. *Econ. Geol.,* **69,** 1191-1206.

Christofferson, H.C., Wallin, B., Selkman, S. & Rickard, D.T. (1979) Mineralisation controls in the sandstone lead-zinc deposits at Vassbo, Sweden. *Econ. Geol.,* **74,** 1239-1249.

Clemmey, H. (1978) Implications of recent copper sulphide placer concentration in Chile. *Trans. Inst. Min. & Met.*, **87,** B 32 (Abs).

Clemmey, H. & Badham, J.P.N. (In Prep.) A review of the geological evidence for the development of the Earth's Atmosphere.

Cloud, P. (1972) A working model of the Primitive Earth. *Amer. J. Sci.*, **272,** 537.

Cloud, P. (1976) Beginnings of biospheric evolution and their biogeochemical consequences. *Palaeobiology*, **2,** 351-387.

Coleman, M.L. (1977) The application sulphur isotope geochemistry to oil, gas and ore studies: a review. In *Forum on Oil and Ore in Sediments*. (Ed. P. Garrard) Imperial College, pp. 53-90.

Cotton, R.E. (1965) The H.Y.C. Lead-zinc-silver ore deposit, McArthur River. In *Geology of Australian ore deposits*. (Ed. J. McAndrew) Comm. Min. Met. Congress, pp. 197-200.

Crowell, J.C. (1974) Origin of Late Cenozoic Basins in Southern California. In *Tectonics and Sedimentation Soc. Econ. Pal. & Min.*, Sp. Publ. 22, pp. 190-204.

Croxford, N.J.W. & Jephcott, S. (1972) The MacArthur Lead-zinc-silver deposit, NT. *Proc. Aust. Inst. Min. Met.*, **243,** 1-24.

Croxford, N.J.W., Janacek, J., Muir, M.D. & Plumb, K.A. (1973) Micro-organisms of Carpenterian age from the Amelia dolomite, McArthur River, N.T. Australia, *Nature*, **245,** 28-30.

Croxford, N.J.W., Gulson, B.L. & Smith, J.W. (1975) The McArthur River deposit: a review of the current situation. *Min. Dep.*, **10,** 302-304.

Dahlkamp, F.J. (1978) Geological Appraisal of the Key Lake U-Ni Deposits, Northern Saskatchewan. *Econ. Geol.*, **73,** 1430-1449.

Davidson, C.F. (1965) A possible mode of origin of strata-bound copper ores. *Econ. Geol.*, **60,** 942-954.

Dimroth, E. (1972) The Labrador Geosyncline revisited. *Amer. J. Sci.*, **272,** 487-506.

Dimroth, E. (1977a) Facies models—5. Models of physical sedimentation of Iron Formations. *Geosci. Can.*, **4,** 23-30.

Dimroth, E. (1977b) Facies models—6. Diagenetic Facies of Iron Formation. *Geosci. Can.*, **4,** 83-88.

Dimroth, E. & Kimberley, M.M. (1976) Precambrian atmospheric oxygen evidence in the sedimentary distributions of carbon, sulfur, uranium and iron. *Can. J. Earth Sci.*, **13,** 1161-86.

Dixon, C.J. (1979) *Atlas of Economic Mineral Deposits*. Chapman & Hall, 143 pp.

Dorr, J.V.N. (1973) Iron Formation in South America. *Econ. Geol.*, **68,** 1005-1022.

Drever, J.I. (1974) Geochemical model for the origin of Precambrian Banded Iron formations. *Bull. Geol. Soc. Amer.*, **85,** 1099-1106.

Ethier, V.G. & Campbell, F.A. (1977) Tourmaline concentrations in Proterozoic sediments of the southern Cordillera of Canada and their economic significance. *Can. J. Earth Sci.*, **14,** 2348-63.

Freeze, A.C. (1966) On the origin of the Sullivan ore body, Kimberley, B.C. In *Tectonic history and mineral deposits of the Western Cordillera*. Can. Inst. Min. Spec. Vol. 8, p. 263.

Garlick, W.G. (1961) Ore genesis—The syngenetic theory. In *Geology of the North Rhodesian Copperbelt*. (Ed. F. Mendelsohn) MacDonald, pp. 146-165.

Garrard, P. (Ed) (1977) *Forum on oil and ore in sediments*. Imperial College, London 202 pp.

Gordon, A.W. (1975) Distribution by latitude of Phanerozoic evaporite deposits. *J. Geol.*, **83,** 671-684.

Harrison, J.E. (1972) Precambrian Belt Basin of the NW United States: Its Geometry, Sedimentation and Copper Occurrences. *Bull. Geol. Soc. Amer.*, **83,** 1215-1240.

Hegge, M.R. & Rowntree, J.C. (1978) Geologic setting and concepts on the origin of uranium deposits in the East Alligator River Region, N.T. *Econ. Geol.*, **73,** 1420-1429.

Helgesen, H.C. (1970) *A chemical and thermodynamic model of ore deposition in hydrothermal systems*. Geol. Soc. Amer. Spec. Paper 3, pp. 155-186.

Henley, R.W. & Adams, J. (1979) On the evolution of giant gold placers. *Trans. Inst. Min. & Met.*, **88**, B41-50.

Hester, B.W. (1970) Geology and evaluation of placer gold deposits in the Klondike area, Yukon Territory. *Trans. Inst. Min. Metall.*, **79**, B60-67.

Heyl, A.V., Landis, G.P. & Sartman, R.G. (1974) Isotopic evidence for origin of Mississippi Valley-type mineral deposits: a review. *Econ. Geol.*, **69**, 992-1006.

Hitchon, B. (1977) Geochemical links between oil fields and ore deposits in sedimentary rocks. In *Forum on Oil and Ore in sediments* (Ed. P. Garrard) Imperial College, London, pp. 1-35.

Hoffman, P. (1973) Evolution of an early Proterozoic Continental margin: the Coronation geosyncline and associated Aulacogens of the northwest Canadian Shield. In *Evolution of the Pre-Cambrian crust. Phil. Trans. Roy. Soc., Lond.*, **A273**, 447-581.

Holland, H.D. (1973) The oceans: a possible source of iron in iron formations. *Econ. Geol.*, **68**, 1169-1172.

Illies, J.H. (1969) An intercontinental belt of the world rift-system. *Tectonophysics*, **8**, 5-29.

Jackson, S.A. & Beales, F.W. (1967) An aspect of sedimentary basin evolution: the concentration of Mississippi Valley-type ores during late stages of diagenesis. *Bull. Can. Petrol. Geol.*, **15**, 383-433.

James, H.L. (1954) Sedimentary Facies of Iron Formation. *Econ. Geol.*, **49**, 235-293.

Kanasewich, E.R. (1968) Pre Cambrian Rift: genesis of strata bound ore deposits. *Science*, **161**, 1002-1005.

Kent, P.E. (1975) Review of North Sea Basin Development. *J. Geol. Soc., Lond.*, **131**, 435-468.

Kimberley, M.M. (1974) Origin of iron ore by diagenetic replacement of calcareous oolite. *Nature*, **250**, 319-320.

Kirkham, R.V. (1974) A synopsis of Canadian stratiform copper deposits in sedimentary sequences. In *Gisements stratiformes et Provinces Cuprifères.* Proc. Soc. Geol. de Belgique, Liège, pp. 367-382.

Kröner, A. (1977) Precambrian mobile belts of southern and eastern Africa—ancient sutures or sites of ensialic mobility? A case for crustal evolution towards plate tectonics. *Tectonophysics*, **40**, 101-135.

Large, D. (1979) Proximal and distal stratabound ore deposits—a discussion. *Min. Dep.*, **14**, 123-4.

MacQueen, R.W. (1979) Base metal deposits in sedimentary rocks—some approaches. *Geosci. Can.*, **6**, 3-9.

MacQueen, R.W. & Thompson, R.I. (1978) Carbonate-hosted lead-zinc occurrences in northeastern British Columbia with emphasis on the Robb Lake Deposit. *Can. J. Earth Sci.*, **15**, 1737-62.

Maxwell, J.C. (1974) Early Western margin of the United States. In *Geology of Continental Margins* (Eds C.A. Burk & C.L. Drake) Springer Verlag, Berlin, pp. 831-851.

McKenzie, D.P. & Sclater, J.G. (1968) Heat Flow inside the Island Arcs of the N.W. Pacific *J. Geophys. Res.*, **73**, 10.

Mendelsohn, F. (1961) *The Geology of the Northern Rhodesian (Zambian) Copper Belt.* MacDonald, London, 523 pp.

Mitchell, A.H.G. & Garson, M.S. (1976) Mineralisation at Plate Boundaries. *Minl. Sci. Eng.*, **8**, 129-169.

Monger, J.W.H., Souther, J.G. & Gabrielse, H. (1972) Evolution of the Canadian Cordillera: a plate tectonic model. *Amer. J. Sci.*, **272**, 577-602.

Monseur, G. (1977) Mineralisations Cambriennes d'Espagne (Essai de Synthese). *Min. Dep.*, **12**, 331-352.

Morganti, J.D. (1977) Howards Pass: an example of a sedimentary exhalative base metal deposit. *Geol. Assoc. Can. Proc.*, Abs. p. 37.

Morrissey, C.J. (1977) Reflections on ores and the apparent scarcity of oil in Irish Carboniferous sediments. In *Forum on oil and ore in sediments.* (Ed. P. Garrard) Imperial College, London, pp. 147-160.

Morton, R.D. (1974) Sandstone-type uranium deposits in the Proterozoic strata of Northwestern Canada. *IAEA Symposium SM 183-37,* Vienna, pp. 255-271.

Morton, R.D. (1976) The Western and Northern Australian uranium deposits—exploration guides or exploration deterrents for Saskatchewan? In *Uranium in Saskatchewan* (Ed. C. E. Dunn) Sask. Geol. Soc. Sp. Publ. 3, pp. 211-254.

Munha, J. (1979) Blue Amphiboles, Metamorphic Regime and plate tectonic modelling in the Iberian Pyrite Belt. *Cont. Min. Pet.,* **69,** 279-289.

Ncube, A., Zweifel, H. & Amstutz, G.C. (1978) On the occurrences of Bravoite, framboidal pyrite, Marcasite, Pyrrhotite and some possible plant remains in the Laisvall Lead and Zinc Deposit, Sweden. *Nares Jahrbuch Min.,* **132,** 264-283.

Norman, D.I. (1978) Ore deposits related to the Keweenawan Rift. In *Petrology and Geochemistry of Continental Rifts* (Eds E.R. Neumann & I.B. Ramberg) Riedel Publ. Co., Dordrecht, Holland, pp. 245-254.

Olade, M.A. (1976) On the genesis of lead-zinc deposits in Nigeria's Benue Rift (Aulacogen): a re-interpretation. *Niger. J. Min. Geol.,* **13,** 20-27.

Piper, J.D.A. (1976) Palaeomagnetic evidence for a Proterozoic supercontinent. *Phil. Trans. Roy. Soc. A,* **280,** 469-490.

Pretorius, D.A. (1975) The Depositional Environment of the Witwatersrand Goldfields: a chronological review of speculations and observations. *Minls. Sci. Engng.,* **7,** 18-47.

Radtke, A.S. & Russell, M.J. (1978) Relationships between minor elements in Paleozoic sediment rocks and the distribution and chemical compositions of base metal deposits in Ireland. *Geol. Assoc. Can.,* Abs., p. 475.

Rickard, D.T., Williden, M., Marde, Y. & Ryhage, R. (1975) Hydrocarbons associated with lead-zinc ores at Laisvall, Sweden. *Nature,* **255,** 131-2.

Rickard, D.T., Williden, M., Marinder, N.E. & Donnelly, T.H. (1979) Studies on the genesis of the Laisvall Sandstone Lead-Zinc deposit, Sweden. *Econ. Geol.,* **74,** 1255-1285.

Roscoe, S.M. (1973) *The Huronian Supergroup, a Paleoaphebian succession showing evidence of atmospheric evolution.* Geol. Ass. Can. Sp. Paper 12, pp. 31-48.

Russell, M.J. (1975) Lithogeochemical environment of the Tynagh base-metal deposit, Ireland and its bearing on ore deposition. *Trans. Inst. Min. Met.,* **84,** B128-133.

Russell, M.J. (1976) *Incipient plate separation and possible related mineralisation in lands bordering the North Atlantic.* Geol. Assoc. Can. Sp. Paper 14, p. 337.

Russell, M.J. (1978) *The tectonic setting and genesis of the early Carboniferous base metal deposits in Ireland.* Abs. Min. Dep. Studies Group, Dublin.

Russell, M.J. & Smyth, D.K. (1978) Evidence for an early Permian Oceanic rift in the Northern North Atlantic. In *Petrology and Geochemistry of continental rifts* (Eds E.R. Neuman & I.B. Ramberg) Reidel Publ. Co., Dordrecht, Holland, pp. 173-180.

Sangster, D.F. (1979) Plate tectonics and mineral deposits. *Geosci. Can.,* **6,** 185-188.

Sarkar, S.C. (1974) Sulfide mineralisation at Sargipali, Orissa, India. *Econ. Geol.,* **69,** 206.

Sawkins, F.J. (1972) Sulfide ore deposits in relation to Plate Tectonics. *J. Geol.,* **80,** 4, p. 377.

Schlanger, S.O. & Combs, J. (1975) Hydrocarbon potential of marginal basins bounded by an island arc. *Geology,* **3,** 397-401.

Schneider, H.J. & Lehmann, B. (1977) Contribution to a new genetical concept on the Bolivian Tin Province. In *Time and Stratabound ore deposits* (Eds D.D. Klemm & H.J. Schneider) Springer-Verlag, Berlin, pp. 153-167.

Schopf, J.W. (1978) The evolution of the earliest cells. *Sci. Amer.,* **241,** 48-67.

Sharp, J.M. Jr (1978) Energy and Momentum Transport model of the Ouachita Basin and its possible impact in formation of economic mineral deposits. *Econ. Geol.,* **73,** 1057-1068.

Smith, A.G., Briden, J.C. & Drewry, G.E. (1973) Phanerozoic World Maps. In *Organisms and Continents through time* (Ed. N.F. Hughes) Spec. Papers In Pal., pp. 1-42.

Solomon, M. & Walshe, J.J. (1979) The formation of massive sulphide deposits on the sea-floor. *Econ. Geol.,* **74,** 797-813.

Stanton, R.L. (1972) *Ore Petrology.* McGraw-Hill, London, 713 pp.

Stanworth, C.W. (1978) A study of the Sosan Group and its uranium mineralisation, East Arm, Great Slave Lake, Canada. Ph.D. Thesis, University of Southampton, 381 pp.

Stapleton, R.P. (1978) Organic metamorphism and uranium occurrences in the Beaufort Group of South Africa. *Econ. Geol.,* **73,** 283-4.

Taylor, S. & Andrew, C.J. (1978) Silvermines orebodies, Co. Tipperary, Ireland. *Trans. Inst. Min. Met.,* **87,** B111-124.

Thiede, D.S. & Cameron, E.N. (1978) Concentration of heavy metals in the Elk Point Evaporite sequence, Saskatchewan. *Econ. Geol.,* **73,** 405-415.

Towe, K.M. (1978) Early Precambrian oxygen: a case against photosynthesis. *Nature,* **274,** 657-661.

Van de Poll, H.W. (1978) Paleoclimatic control and stratigraphic limits of synsedimentary mineral occurrences in Mississipian-Early Pennsylvanian strata of Eastern Canada. *Econ. Geol.,* **73,** 1069-1081.

Van Eden, J.G. (1974) Depositional and Diagenetic Environment related to Sulfide Mineralisation, Nufulira, Zambia. *Econ. Geol.,* **69,** 59-79.

Van Eden, J.G. (1978) Stratiform copper and zinc mineralisation in the Cretaceous of Angola. *Econ. Geol.,* **73,** 1154-1160.

Walker, R.R. (1977) The geology and Uranium deposits of Proterozoic rocks, Simpson Islands, N.W.T. M.Sc. Thesis, Alberta, 193 pp.

West, I.M., Brandon, A. & Smith, N. (1968) A tidal flat evaporitic facies in the Visean of Ireland. *J. Sedim. Petrol.,* **38,** 1079-1093.

White, W.S. (1971) A Paleohydrologic model for mineralisation of the White Pine copper deposit, Northern Michigan. *Econ. Geol.,* **66,** 1-13.

White, W.S. & Wright, J.C. (1954) The White Pine Copper Deposit, Ontonogan County, Michigan. *Econ. Geol.,* **49,** 675-716.

White, W.S. & Wright, J.C. (1966) Sulfide-mineral zoning in the basal Nonesuch shale, northern Michigan. *Econ. Geol.,* **61,** 1171-1190.

Williams, N. (1978) Studies of the Base Metal Deposits at McArthur River, Northern Territory, Australia. I. The Cooley & Ridge Deposits. II. The Sulphide-S and Organic-C Relationships of the Concordant deposits and their significance. *Econ. Geol.,* **73,** 1005-1035; 1036-1056.

Windley, B.F. (1977) *The Evolving Continents.* John Wiley & Sons, 373 pp.

Yeo, G.M. (1978) *Iron Formation in the Rapitan Group, Mackenzie Mountains.* Minl. Industry Rept. EGS D.I.A.N.D.

Young, G.M. (1976) Iron-formation and glaciogenic rocks of the Rapitan Group, NWT, Canada. *Precamb. Res.,* **3,** 137-158.

Zuffardi, P. (1967) Sedimentary ores of the Middle Silurian in Sardinia. In *Sedimentary Ores: Ancient and Modern.* (Ed. T. James) University of Leicester, pp. 207-239.

Chapter 9 Palaeoclimatic Considerations and General Conclusions

D.H. TARLING

9.1 Introduction

In the preceding chapters individual authors have considered specific types of economic resources within a general, but not necessarily specifically, geotectonic framework. It is clear that geotectonic processes are of varying relevance to the actual prediction of the occurrence of new resources, although such concepts aid an understanding of the processes by which such resources have developed. There have also been indications, particularly in terms of the organic materials, coal and hydrocarbons, and of the 'sedimentary' ores, that palaeoclimatic conditions are also of direct relevance. As the ancient climates differed from those of today as a result of different latitudinal palaeogeographic conditions arising from geotectonic process, it seems appropriate to consider the importance of palaeoclimates separately, before an overall assessment is made.

9.2 Palaeoclimatic factors

It is difficult to distinguish between world-wide climatic changes, changes due to the motion of continental blocks into different latitudes and those due to changes in the local palaeogeography. There is an increasing consensus that the climatic fluctuations of the last million years or so are directly attributable to complex cyclical variations in the Earth-Sun distance (Milankovitch, 1938; Evans, 1971) and similar such variations must have occurred throughout geological time. It is also known that the solar constant is not, in fact, constant, but the maximum possible amplitudes and periodicities of such changes in the total solar radiation are unknown. It is also possible that the total radiation reaching the Earth may have differed in quality or quantity if, for example, the solar system passed through interstellar dust clouds (Williams, 1975a). Williams (1975b) has also suggested that the inclination of the Earth's rotational axis may have changed with time, and that the world-wide Eocambrian glaciations (Harland & Herod, 1975) are attributed to a time when the Earth's rotational axis lay within the plan of the ecliptic.

There is clear evidence that the present climatic regime of the Earth is abnormal when considered on a geological time scale. The Jurassic and Cretaceous periods (c. 100-200 million years ago) are widely recognised as being characterised by world-wide ameliorable climates and polar ice sheets appear to characterise only intervals of a few million years at widely separated intervals (Tarling, 1978). Williams (1975b) suggested that these intervals are some 155 million years and could be related to the rotation period of the solar system around the galaxy, 136 million years.

The geological observations do not, in fact, confirm such a periodicity but there does appear to be a correlation with the existence of specific polar palaeogeographic conditions which result in an emphasis of the effects of the Milankovitch cycles (Tarling, 1978). Such a correlation, with the agreement between palaeoclimate and palaeolatitudes, would also suggest that changes in the inclination of the Earth's axis of rotation do not account for the Eocambrian ice age(s). Indeed, it seems likely that most apparently gross changes in the world's climatic patterns can be explained in terms of palaeogeographic changes resulting from the motion of the continental blocks into different latitudes and configurations. The absence of polar ice caps, for example, would mean that the circulation system of the atmosphere would be drastically different from when the ice caps existed. The total radiation falling on the summer pole would then be greater than that falling on the equator so that the atmospheric Hadley cell would rise at the pole during the summer, creating a zone of low pressure, probably associated with high rainfall on its periphery. Polar areas would, of course, still be cold during the winter and would then be characterised by high pressures, but the presence of warmer waters would alleviate the temperature extremes even during the middle of winter. The present atmospheric circulation is, of course, complex and difficult to simulate today even with known parameters, so it is even more difficult to determine its pattern in the past. The palaeolatitudinal distribution of red beds, evaporites and thick carbonates seem to confirm the presence of an equatorial rain belt during most of the Earth's history (Tarling, 1971). The main effect of the reduction in albedo in polar regions, therefore, appears to have been to produce somewhat broader climatic belts with a generally more equable climate.

Such a broad generalisation about the nature of past latitudinal zoning must be tempered by the regional and local effects of palaeogeography. The generation of mountain belts by geotectonic processes may result in the formation of rain-shadow areas, monsoonal conditions, and so forth. Nonetheless, as the distributions of the main continental blocks become better established for different periods, it becomes increasingly possible to determine the distribution of land and sea relative to the latitudinal zones and hence predict, to a first order, areas of high and low atmospheric pressure, the probable oceanic circulation system and hence assess regional climatic regimes at different times. In such a context, it is therefore possible to consider specific sedimentary facies in terms of the economic importance and probable palaeoclimatic and palaeolatitudinal significance, together with their economic significance (Tarling, 1977).

EQUATORIAL DEPOSITS

Equatorial deposits are likely to be organic rich, reflecting the higher biological productivity of such regions, and the high rainfall will obviously contribute to the formation of peat and eventually coal (Chapter 5). Low-latitude thick coals do, in fact, appear to be confined to within some 10° of the palaeoequator (Tarling, 1971). The intensity of chemical weathering under such conditions is also high, promoting the formation of residual ores, such as bauxites and laterites, and is a major factor in promoting supergene enrichment of, for example, the banded iron formation deposits (Chapter 8). Conversely, the passage of a continent through such environments is also likely to have lead to the removal of chemical precipitates from the superficial rocks. Obviously such processes are not confined to equatorial latitudes and any areas with a moderate rainfall in low latitudes are likely to have enhanced chemical activity resulting, for example, in the formation of kaolin from granites, etc. In this connection, the actual distribution of the equatorial rain belt is likely to very strongly influenced by the distribution of land and sea masses, with the generation of monsoonal conditions under specific palaeogeographic conditions.

EVAPORITIC DEPOSITS

Evaporitic deposits are of prime importance in the formation of hydrocarbon resources as their ability to flow on very short geological time scales makes them an ideal seal, preventing the migration of both liquids and gases. In addition their low density means that they tend to rise as diapirs through the sediments that bury them. This rise tends to dome the overlying sediments into anticlinal traps, although the salt can pierce the dome, forming surface salt glaciers in areas of very low rainfall. The diapirs also tend to spread out when nearing the surface so that hydrocarbons migrating along the tilted country rocks also become sealed in stratigraphic traps. Although salt is itself deficient in organic materials, its low latitude association with organic-rich deposits, such as algal mats in sabkha environments, means that it is often closely associated with hydrocarbon source rocks (Tarling, 1973a, 1977). A

Chapter 9

further important factor, in relationship to hydrocarbon accumulations, is that salt has a high thermal conductivity which can operate to concentrate heat in overlying rocks and transport it from lower rocks. This means that the presence of salt deposits tends to even out the local temperatures thereby improving the regional maturation and tend to suppress hydrocarbon thermal destruction. In terms of economic minerals, anhydrite is of fundamental importance to the fertiliser industry, particularly superphosphates and potash fertilisers, and anhydrite, in addition to fertiliser applications, is the main source of plaster. Such evaporitic deposits are therefore of major importance in their own right, but the presence of at least some evaporites are useful in providing a source for the formation of the metallic chloride complexes associated with some ground water-derived metallic ores (Section 8.3). Although major deposits are not essential for this purpose and sea water is itself available in the marine environment (Chapter 6), the fluid inclusions within Mississippi Valley-type ores are usually considered to be of sabkha brines. Furthermore, the sulphur isotopes present in many of these types of deposit are of shallow origin and usually thought to be derived from calcium-sulphate deposits.

The presence of thick evaporitic sequences have long been used as indicators of low latitudes and palaeomagnetically determined palaeolatitudes (Fig. 9.1) confirms that most major deposits have formed

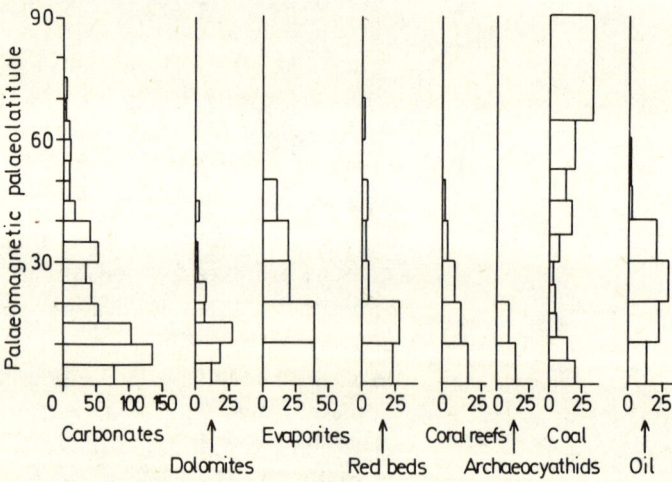

Fig. 9.1 The palaeomagnetically determined palaeolatitudinal distribution of some geological materials. The distribution of these materials is based on palaeomagnetic determinations of the inclination of magnetisation in them, or in nearby rocks of similar age. Although the reliability of some of the data is in doubt, there is a clear correspondence between the determined palaeolatitudinal distribution and the palaeoclimatic conditions that appear to be associated with their optimal formation. Stratiform ore bodies are often associated with reefs, thick carbonate sequences and evaporites and hence these also show a strong palaeolatitudinal control. (After Tarling, 1971.)

within 40° of the palaeoequator (Irving & Briden, 1962; Tarling, 1971, 1973a, 1977). The reasons for this low-latitude association are, however, relevant to the close association between hydrocarbon deposits and evaporites—approximately half of known reserves occur with evaporite-associated facies (Moody, 1975). Virtually all known salt deposits have

Palaeoclimatic Considerations and General Conclusions

not arisen by the simple evaporation of sea water so quite specific palaeo-environments are required in areas of high evaporation (Section 2.2). Two main conditions have long been proposed, the silled basin and the sabkha (Kinsman, 1969), but both require these very specific controls to be maintained for a long period if thick deposits (3 km thicknesses are known in Texas) are to form (Tarling, 1973a). The silled basin environment requires the position of the sill to be maintained relative to sea-level in an area where the basement is sinking by possibly several kilometres. Similarly, the sabkha environment requires the shoreline to be maintained in roughly the same location as similar basement depression takes place. It is difficult for such structural controls to be maintained without the assistance of organic reef growth that can maintain its position relative to sea-level for prolonged periods even when the sinking is rapid. Such conditions can be maintained for short periods by structural geotectonic factors. In the case of the Cretaceous evaporites in the South Atlantic, the Angola and Brazilian Basins, the silled basin condition was maintained by the very slow subsidence rates of the Walvis-Rio Grande Rise in contrast to the ocean basins to the north and south (Argentine and Cape Basins) where no hot spots existed. In these basins the ocean floor was already close or slightly below sea-level shortly after the start of opening of the South Atlantic (Tarling, 1980). However, the importance of palaeolatitude is again indicated as identical geotectonic conditions occurred south of the Walvis-Rio Grande Rise, but this area lay at 50-60°S, while the area to the north lay at 20-30°S. The prolonged evaporitic sequences of the Lower Palaeozoic of Canada, Upper Palaeozoic of the United States and Europe are, however, predominantly of sabkha origin and again strongly constrained by the changing palaeolatitudes at this time.

BIOHERMAL REEFS

Biohermal reefs are significant reservoirs for hydrocarbons or metallic fluids as they tend to be permeable and form tectonic highs that may be sealed by evaporites. The more general significance of such reefs is their control of local palaeogeographic conditions (as discussed in relationship to evaporites and carbonates). The growth of present day reef corals within 30° of the equator does not necessarily mean that Palaeozoic corals had identical limitations, although their palaeolatitudinal distribution within 30° of the palaeoequator (Fig. 9.1) would suggest that similar factors were critical, namely warm clear shallow marine waters, such as occur in low latitudes today.

Chapter 9

CARBONATES

Carbonates have an intrinsic economic importance as they are themselves valuable sources for aggregate. In slightly impure forms they are also the source for cement. The association of biogenic limestones with high organic productivity may partially explain why approximately half of the world's known supplies of hydrocarbons are in carbonate reservoirs (although such a figure is strongly biased by the Middle Eastern oilfields). The presence and age of dolomite is particularly important as its porosity and general high permeability mean that they are of prime importance for migration and accumulation of hydrocarbons or metal-rich ground waters. The presence in carbonates of organic material, and hence sulphur, also provides a reactant for the metallic fluids carried in either ground waters or hydrothermal systems, while the Eh-pH conditions appear to favour the deposition of lead and zinc sulphides emanating from either igneous (Chapter 6 & 7) or ground-water sources (Chapter 8). The factors that control the location of carbonate deposits at different times are therefore of crucial importance in evaluating the economic potential of an area for hydrocarbons, metals and the carbonates themselves. Both chemical and organic carbonates are strongly confined to low to intermediate latitudes because of the dependence of the rate of $CaCO_3$ deposition on temperature and salinity. Virtually all thick dolomites are confined to within 25° of the palaeoequator and massive limestones form between 5 and 40° of the palaeoequator (Fig. 9.1; Tarling, 1971). Biogenic carbonate concentrations will be further enhanced in low latitudes by the generally higher organic productivity in such latitudes, but high productivity can also occur in somewhat higher latitudes if there is a plentiful supply of nutrients. The deposition of pure Cretaceous chalk in northwestern Europe took place at some 40°N, but reflected the mixing of oceanic waters as the Labrador Sea opened to supply Arctic waters into the proto-Gulf Stream (Hart & Tarling, 1974). Similarly, upwelling bottom waters, usually on the western coasts of continents, can give rise to prolific organic growth (Section 2.2) that can form thick carbonate depositions somewhat outside the tropics. The presence of quantities of terrestrial detritus, however, not only dilutes and pollutes the carbonate deposits, but tends to inhibit calcareous-secreting organisms. The presence of high rainfall in equatorial regions may therefore account for the suppression of thick carbonates within 5° of the Equator and, conversely, it seems likely that optimum areas for biogenic carbonates will occur immediately offshore from desert areas.

The accumulation of thick and laterally extensive carbonates also requires the subsidence of land, relative to sea-level, at a rate not exceeding the rate of deposition of the carbonates. The most extensive lateral subsidence is likely to be associated with the trailing edges of rifted continental margins (Sections 3.1). The rate of deposition may well be enhanced if the carbonate platforms are also associated with reefs between the sinking basins and the deep oceans. Such restricted basins are likely to have greater nutrient concentration, hence greater organic productivity and a greater ability to maintain deposition in areas of rapid basement sinking. Such conditions are also likely to be conducive to the rapid development of algal limestones (Tarling, 1973a, 1977). It seems probable that these are often a major source for hydrocarbons. In the Belle River Mills gas field in Michigan, for example, algal limestones are the only possible source for the hydrocarbons in the reef reservoir (Gill, 1973). Similarly, oil seepages occur from the Carboniferous algal limestones in the Paradox Basin, Utah, and in the internal lakes of the Green River shales of Wyoming and Idaho. Indeed, organic content of algal limestones may well be a major source for the Middle East oilfields, for example the Jurassic of Abu Dhabi. Somewhat less laterally extensive, and usually more restricted in duration, will be carbonates forming in basins on the leading edges of continents. In such areas, the tectonic regimes are variable in extent and duration, particularly if associated with an active subduction zone. However, such leading edges may be protected from major tectonic disturbance by the presence of an offshore island arc, such as is the case of northwestern Australia today.

SANDSTONES

Sandstones are a major reservoir rock for hydrocarbons, containing approximately half of the world's known reserves. The sandstones that have formed as dunes tend to have the greatest hydrocarbon content as they tend to be highly permeable, thus providing migration channels and reservoirs. Metallic deposits are sometimes associated with sandstones, partially the uranium-vanadium association with the so-called red beds (Section 8.3) that have formed in hot desert conditions.

These sediments can, of course, form in any latitude and thick sandstone deposits can be regarded, in some senses, as indicating the absence of clay-forming conditions. As such, sands are usually strongly associated with glacial activity, but these are generally reworked before undergoing diagenesis. In general, sandstones are predominantly formed in low to intermediate latitudes, often in close association with hot desert conditions in which hydrous clay formation tends to be suppressed. A further constraint is, of course, the availability of continental sources with an obvious compromise required between the availability of elevated land for erosion and sufficient river length to give a significant grain-size sorting. One major problem in the evaluation of the palaeoclimate indicated by sandstones is that the reddening usually attributed to sandstones that have formed in hot desert environments may often have developed many millions of years after their deposits. The reddening can thus be a misleading indicator of the likely palaeolatitude at which such sandstones were deposited. As with shales, the conditions of formations of thick sandstone sequences are likely to have changed dramatically as the terrestrial surface began to become vegetated in Devonian times, and again with the advent of gymnosperms at the end of the Cretaceous.

SHALES

Shales are of major importance as potential source rocks for hydrocarbons, especially those of marine origin. As organic productivity is enhanced at low latitudes and where nutrients are available, the association of virtually all major oil- and gas-fields with a palaeogeographic distribution within 5 to 40° of the palaeoequator (Irving & Gaskell, 1964; Deutsch, 1965; Tarling, 1971) suggests that this biological activity is of prime importance, the few exceptions apparently being mainly associated with deltaic conditions but still at intermediate latitudes. In areas of rapid deposition, such as deltas, the clays may also become mobilised, forming clay diapirs analogous to those associated with salt, but these clays also have important effects as they act as a thermal blanket as well as forming seals against further migration of fluids or gases. The high chemical activity of shales also makes them sites for the concentration of metals from passing fluids, particularly during their deposition, while their organic content may provide sulphur for reactions with soluble metallic chloride complexes, thereby causing metallic sulphide deposition. However, their impermeability means that most of such activity is restricted to isolated clays within host rocks, such as carbonates and sandstones.

GLACIAL DEPOSITS

Glacial deposits provide major sources of sand and gravels for construction and glass manufacture. These must be unconsolidated for most economic use and are thus mainly associated with the Quaternary ice age. While the cause of this series of ice advances and retreats may reflect a combination of astronomical and geophysical changes, a fuller understanding of the origin of such ice ages is unlikely to improve methods significantly for the location of new reserves of these important building materials. The knowledge of the origin of older ice ages is likely to be even less useful as these tend to be more consolidated and thus uneconomic. Similarly, glaciation arising from elevation by geotectonic processes is of little direct value as only recent deposits will be useful. However, the emphasis on mechanical erosion and transport, rather than chemical activity, means that placer deposits can sometimes be associated with such processes where a pre-existing ore deposit has been eroded, but such concentrations will necessarily be essentially of a mechanical nature, such as the copper-sulphide deposits of Kennecott, Alaska.

The establishment of high latitude rainfall belts shortly after the cessation of Lower Permian (Sakmarian) glaciation of Gondwanaland and was obviously critical in the formation of the Gondwanan high latitude coals (Chapter 5) most of which formed at latitudes greater than 60°. On this basis, it is important to establish the actual constitution of Gondwanaland at that time as similar conditions could thus be predicted to have existed in such fragments. However, the areas of uncertainty mostly lie in the Far East where increasing supplies of brown coals of more recent age are already being discovered.

9.3 General conclusions

Although the authors were specifically requested to consider aspects in which they considered geotectonic processes to be irrelevant, there has possibly been a tendency to emphasise those aspects that are more directly related to these processes. The main consensus, however, is that while geotectonic processes are sometimes of direct significance, for example in terms of the generation of ore deposits at ridges and subduction zones, their main importance is in the improved knowledge of the possible modes of origin of these resources. Such a conclusion seems to follow from the observation that the formation of economic reserves occurs on a scale that is much smaller than that of geotectonic processes. The influence of geotectonic processes can be considered in terms of their influence on heat-flow distributions, the structural situations in terms of fracturing, rifts, etc., and the palaeogeographic conditions associated with different locations within a plate framework. Naturally the fact that plate tectonics involves a mobile Earth's surface must also mean that individual situations change with time so that the actual sequence of geotectonically controlled environments may be as important as the occurrence of one particular environment at one particular time. For example, hydrocarbons may well accumulate during the early rifting of an ocean, but this ocean must also ultimately close. If these hydrocarbons were able to survive until the closure, which is itself unlikely, they would certainly be destroyed during the collision, unless the colliding edges were sufficiently irregular that odd small quasi-oceanic basins remained, for example, the Caspian Sea is in part a relict from the closure of the Tethys ocean. Similarly, there is certainly no consensus on the relevance of plate tectonic processes in the past and, if they occurred, their nature and effect was likely to have been drastically different at earlier times in the Earth's history (Chapter 1).

HEAT-FLOW AND VOLCANISM

The fact that present day plate boundaries are defined by the occurrence of earthquakes and, to a lesser extent, volcanic activity immediately indicates the concentration of heat-flow and tectonic activity at such boundaries. The significance of volcanic activity at oceanic ridges and subduction zones has been described in some detail (Chapters 6 and 7). The maturation, migration and destructive effects of high heat-flow associated with the oceanic ridges and with the volcanic activity bordering a subduction zone have also been considered in terms of hydrocarbon accumulations (Chapters 3 and 4). In all these cases, the relationship with hydrocarbon accumulations and ore deposits is clear. However, only some 80% of volcanic activity is directly associated with plate boundaries. In a mobile Earth, such as required by plate tectonics, it seems evident that the remaining fifth of volcanic activity must be explicable in relationship to the revised view of the nature of the Earth's upper levels, if not to geotectonic processes directly.

One obvious situation is that of the kimberlites and carbonatites that appear to be associated respectively with Archaean cratons and the early stages of rifting. The kimberlites clearly indicate that the continental lithosphere is some 180-200 km (Section 1.3), but the reason why kimberlite activity should be initiated at different times is by no means clear. It is difficult to see how thermal diapirs can arise within the mantle itself as its rigidity would mean that any such motions would be rapidly transferred to the rest of the mantle, thus inhibiting rapid diapiric rise (Runcorn, 1974). It is conceivable that heat could accumulate beneath a continental lithosphere, but it seems more likely that major lower continental lithosphere activity would be generated if volatiles were added to it. The addition of volatiles would reduce the melting temperature of the mantle rocks which may then become molten at ambient temperature. Once fluids have been generated, these would then have a much lower density and would rise to the surface, loosing heat to the surrounding rocks, but also rising to lower pressure areas and thus into areas in which their melting points would be somewhat reduced. In the case of kimberlites, it seems probable that the initial motion would be sufficient to initiate carbon-dioxide escape which, once initiated, would cause rapid rise and explosive loss of other volatiles from the ascending magma. The addition of volatiles to the base of the continental lithosphere could arise from local inhomogeneities within the circulating mantle convection currents, but would more readily arise by the subduction of oceanic lithosphere in the nearby area, or possibly by delayed dehydration processes in marginal lithosphere (Section 1.3).

Carbonatites could similarly be generated during the initiation of subcontinental lithosphere convective uprise, resulting in the addition of some volatiles giving preliminary carbonatitic activity before mantle convection motions had penetrated to shallow depths from which more tholeitic magmas could be generated. While the precise origin of these types of igneous rock is certainly not clear, the plate tectonic models at least provide a conceptual framework for the origin of such materials. More importantly, it does suggest fairly specific environments within which such volcanic activity can be expected to have occurred. Such materials are particularly valuable for many of the rarer elements, such as diamonds, strontium, the rare earth elements in general, etc.

Such a model does not seem to be quite as appropriate for some other intraplate volcanic activity. Some volcanoes may be associated with the extension of oceanic fracture zones into continental regimes, but while such an explanation cannot be excluded, for example, for the Hoggar Massif in the central Sahara, it is still difficult to see how the sub-continental lithospheric processes would have been generated. Similarly it is not clear why apparently sporadic doming, sometimes accompanied by volcanic activity, should occur at all. There seems to be indications of doming of various parts of the African lithosphere, some of which have partially linked to form the East African Rift system, while others appear to have domed and then collapsed. The Rhine Graben, for example, has volcanism and residual high heat-flow, but there is no obvious reason for its location or apparent history, although probably related to the Alpine orogenesis. One possible explanation is that most continental lithospheres already have a distinct tectonic fracture pattern, often of Precambrian age. The imposition of stresses from below and from different sides could cause some degree of fracture opening, releasing the pressure on underlying rocks which then approach their melting point or begin to loose volatiles that reduce the melting point elsewhere. [This model is analogous to that proposed to account for the uneven lateral distribution of ore deposits and volcanism bordering a subduction zone, such as the Peru-Chile Trench (Tarling, 1973b), although the interaction, in that case, was between the stress directions, the fracture patterns and the stress direction determined by the thickness of oceanic lithosphere being subducted.]

Chapter 9

STRUCTURAL CONTROLS

The association with plate boundaries and earthquakes clearly indicates the increase in tectonic activity that can be expected near boundaries. The importance of this in terms of hydrocarbons and ore deposits has already been made evident (Chapters 3 and 7). As in the case of heat-flow, however, some one fifth of current earthquake activity occurs away from accepted plate boundaries. The pattern and fault plane solutions for this activity also seems to indicate that a fairly uniform stress field is operating within continents (Richardson *et al.*, 1979). If sub-lithosphere mantle drag is the main driving force (Section 1.5), then there will be a relationship between the direction of this stress field and any pre-existing fracture pattern in the crustal rocks. However, the lower crust and underlying mantle lithosphere appear to be able to yield sufficiently to prevent sufficient stress accumulation to result in eventual brittle failure as virtually all earthquakes occurring away from plate boundaries within a continent occur in the upper 10-12 km. Such a model would be of importance in determining the orientation of open or closed fracture systems at different times, and hence the availability of channels for the migration of ground waters containing metallic solutes or hydrocarbons. However, the real situation is clearly much more complex. Differential rates of cooling, for example, would result in stress accumulation between adjacent units of the continent that had been subjected to orogenic or thermal activity at different times. The origin of such activity is clearly of major importance in understanding the development of intracontinental basins. It has been suggested that the sinking of circular basins, such as the Michigan Basin, can be explained simply in terms of a cooling of the lithosphere with associated contraction and isostatic adjustment (Turcotte & Ahern, 1977; Beaumont, 1978). This again raises the problem of how localised heating occurred in the first place. Nonetheless, the recognition of the cooling and contractional rates in the continents and oceans is of fundamental importance in understanding how such basin subsidence takes place and how fracture systems surrounding such basins may have been active at specific times. The basin itself would, of course, be a potential area for hydrocarbons (Sections 3.1) and the fracture systems would provide conduits for metallogenic fluids (Section 9.3).

At this moment, it is difficult to understand the origin of such tectonic activity either without or within a plate tectonic framework, but it seems clear that as an understanding of geotectonic processes, clearly associated with mantle convective processes, must eventually provide an understanding of the origin of intraplate features of major economic importance.

Palaeoclimatic Considerations and General Conclusions

PALAEOGEOGRAPHY AND PALAEOCLIMATE

The significance of palaeoclimatic factors has been discussed earlier and the realtionship between palaeogeography and geotectonic processes are fairly clear. Plate boundaries, with high tectonic activity and heat-flow, tend to be areas of high topography and rapid erosion, etc., while the continental side of a subduction zone is likely to be depressed in response to the loading on its edge, for example the eastern edge of the Andes, in addition to the effect of mantle and intraplate forces. Such basins are likely areas for the accumulation of placer deposits (Section 8.3) and are also potentially important for both source rocks and traps for hydrocarbons migrating away from the bordering higher geothermal gradients (Chapters 2 and 3). Away from these areas, mechanical erosion will be low as topography is likely to be subdued, with most changes occurring in response to changes of sea-level. The trailing edges of continents are likely to be downwarped if they are rigidly coupled to the cooling and contracting denser oceanic lithosphere contiguous with their edges. Such edges are also likely to be strongly fractured during the earlier rifting processes of ocean formation and thus provide areas for hydrocarbon accumulation, migration and accumulation (Chapter 3).

A major difficulty in assessing possible palaeoclimatic relationships is the paucity and reliability of much of the palaeomagnetic data for the Palaeozoic and earlier times (Tarling, 1979). However, the palaeoclimatic effect of local palaeogeographic distributions may often mean that the precise location of the palaeolatitude is unnecessary in any evaluation of the hydrocarbon or metallic ore resource of any particular area at any particular time. It may be sufficient to assess that it was then in high, intermediate or low latitudes.

CONCLUSIONS

This book was not devised in order to 'explain' economic geology in terms of relatively new concepts of the Earth's development. It should be remembered that an explanation 'in terms of plate tectonics' may be meaningless at a time when there is still no satisfactory physical definition of a tectonic plate. Similarly, the advent of the plate tectonic concept does not invalidate all previous geological observations and considerations. Oil will still be found by wild-catting for some time to come! However, the model does provide a framework within which it is much easier to understand the interrelationship of all of the earth sciences. Economic geology is important, not merely in meeting Man's future needs, but as an integration of the geological disciplines and, as such, is necessarily intimately involved with any ideas that provide a clearer synthesis of our understanding of Earth processes. It is only with such an understanding that techniques can be developed and improved in order to both discover new reserves and, possibly more importantly, discover more effective ways of developing known reserves.

References

Beaumont, C. (1978) The evolution of sedimentary basins on a viscoelastic lithosphere : theory and examples. *Geophys. J. R. astr. Soc.,* **55,** 471-497.

Deutsch, E.R. (1965) The Paleolatitude of Tertiary Oil Fields. *J. Geophys. Res.,* **70,** 5193-5203.

Evans, P. (1971) Towards a Pleistocene Time-scale. In *The Phanerozoic Time-scale—a supplement.* Geol. Soc., London. Spec. Publ. 5, pp. 123-356.

Gill, D. (1973) Stratigraphy facies, evolution and diagenesis of productive Niagaran Guelph Reefs and Cayngan Sabkha deposits, the Belle River Mills Field, Michigan Basin. Ph.D. thesis, Univ. Michigan, Ann Arbor, 282 pp.

Harland, W.B. & Herod, K.N. (1975) Glaciations through time. In *Ice Ages : Ancient and Modern* (Ed. A. E. Wright & F. Moseley) Seel House Press, Liverpool, pp. 189-216.

Hart, M.B. & Tarling, D.H. (1974) Cenomanian Palaeogeography of the North Atlantic and possible Mid-Cenomanian Eustatic Movements and their implications. *Palaeogeog., Palaeoclimate., Palaeoecol.,* **15,** 95-108.

Irving, E., & Briden, J.C. (1962) Palaeolatitude of Evaporite Deposits, *Nature,* **196,** 425-428.

Irving, E. & Gaskell, T.F. (1962) The Palaeogeographic Latitude of Oil Fields. *Geog. J.,* **7,** 54-64.

Kinsman, D.J.J. (1969) Modes of formation, sedimentary associations, and diagnostic features of shallow-water and supratidal evaporites. *Amer. Assoc. Petrol. Geol. Bull.,* **53,** 830-840.

Milankovitch, M. (1938) Neue Engebuisse der astronomischen Theorie der Klimaschwankungen. *Bull. Acad. Sci. math. nat. Belgr. A,* **4,** 1-41.

Moody, J.D. (1975) Distribution and geological characteristics of giant oil fields. In *Petroleum and Global Tectonics* (Eds A.G. Fischer & S. Judson) Princeton Univ. Press, Princeton, pp. 307-320.

Richardson, R.M., Solomon, R.C. & Sleep, N.H. (1979) Tectonic stress in the plates. *Revs. Geophys. Space Phys.,* **17,** 981-1019.

Runcorn, S.K. (1974) On the forces not moving lithospheric plates. *Tectonophys.,* **21,** 197-202.

Tarling, D.H. (1971) *Principles and Applications of Palaeomagnetism.* Chapman & Hall, London, 164 pp.

Tarling, D.H. (1973a) Continental Drift and Reserves of Oil and Natural Gas. *Nature,* **243,** 277-279.

Tarling, D.H. (1973b) Metallic Ore Deposits and Continental Drift. *Nature,* **43,** 193-196.

Tarling, D.H. (1977) Some Economic implications of Continental Drift. *Naturwissenschaften,* **64,** 16-22.

Tarling, D.H. (1978) Geological-Geophysical Aspects of Ice Ages. In *Climatic Change* (Ed. J. Gribbin) Cambridge University Press, Cambridge, pp. 3-24.

Tarling, D.H. (1979) Palaeomagnetic Reconstructions and the Variscan Orogeny. *Proc. Ussher Soc.,* **4,** 233-260.

Tarling, D.H. (1980) The Geologic Evolution of South America during the last 200 million years. In *Evolutionary Biology of the New World Monkeys and Continental Drift* (Eds R.L. Ciochon & A.B. Chiarelli) Plenum, N.Y.

Turcotte, D.L. & Ahern, J.L. (1977) On the thermal and subsidence history of sedimentary basins. *J. Geophys. Res.,* **82,** 3762-66.

Williams, G.E. (1975a) Possible relation between periodic glaciation and the flexure of the galaxy. *Earth Planet. Sci. Letters,* **26,** 361-369.

Williams, G.E. (1975b) Late Precambrian glacial climate and the Earth's obliquity. *Geol. Mag.,* **112,** 441-544.

Index

Page numbers in *italics* refer to main references. Page numbers in **bold** refer to pages on which illustrations appear.

Abu Dhabi, 199
Acadian orogeny, 107,108
Adamellites, 141
Aden, Gulf of, 16,129
Adiabatic gradient, 120
Aegir marine band, 111
Aegiranum marine band, 111
Aeolian sands, 78
Africa, 13,14,15,109,113
   Central, Copperbelt, 161
   East, rift valleys, 53,203
   North, 64
   Southern, 14
   West, 15,114
Alaska, 33,57,59,**60**,**83**,152,181
   Cook inlet of, 59,**60**,84,**85**
   Gulf of, **58**
Alberta, 51,109,169
Aleutian Arc, 57
Aleutian Trench, 11
Algae, 37,100,161,174,179,195,199
Alleghanian orogeny, 113
Alps, 3,10,69
Amphibolite, 22
Amposta oilfield, 91
Andaman Sea, 60,61,**62**
Andes, 15,60,**61**,93,137,205
Angaraland, 109,114
Angiosperms, 99
Angola, 54,66,160,**162**,197
Anhydrite, 77,196
Anoxia, water, 37,40,41
Antarctica, 13,14,15
Anthracite, **102**,103,104,110,115
Anthracite coalfield, 113
Anticlinal traps, 195
Anticlines, 89
   compressional, 84-85
   non-compressional, 86,87
   'roll-over', 69,87
Antimony, 174,176,180,181
Appalachian Basin, 111,113
Appalachian mountains, 113,137
   Canadian, copper in, **162**

Aqueous flushing, 46
Arabia, **68**, 70
Arabian Shield, 75
Archaean Period, 11,12,20,21,22,23,150
   sedimentary ore deposits, 150,154,155, 158,177
Arctic Ocean, 16,18,**32**,33,198
Argentine, 197
Argillaceous rocks, 45,47
Argyll fields, 55,**56**
Aromatics, 34
Arsenic, 143,174,176,177,180,181
Ash, mineral, 110
Ash content, 103
Asia, 3,143
Asthenosphere, **2**,**3**,20,119,**120**,**126**
Atacama desert, 160,**162**,164
Athabasca sandstone, 167
Atlantic Ocean, 7,14,18,53,92,123
   central, 15
   North, 5,55
   ridge, 7,129
   South, **5**,8,15,166,172,197
Atlantis Deep, 127,**128**
Atmosphere, 151,157,158,194
Augila Field, Libya, 83
Auk Field, 55
Aulacogen, **9**,54,78,114,173,180,181
Australia, 12,13,14,15,92,167,199
   coal, 101,111,114
Azores, 6

Back-arc basin, **57**,*59-61*,88
Back-arc magmatic belts, *136-137*,140
Bacteria, 34,100,133
Baffin Bay, 15
Baltic Sea, 39
Baltic Shield, 112
Banded Iron Formation, 155,158,195
Barite, 140,168,174,181
Barium, 172,173,174,177,182,184
Barytes, 140,168,174,181
Basalts, 15,121,122,123,**126**

Basin and Range Province, USA, 4
Basins, 169
   classification, 51
   marine, 38-39,144
   see also specific basin types
Batten Trough, **178**
Bauxites, 195
Beaufort Basin, South Africa, 167
Belgium, 112,176
Belle River Mills, 199
Beltian Delta, 180
Benioff zone, 9,10,22,135,136,137,143
Benue Trough, 180
Billiton, 140
Biosphere, 151
Bioturbation, 132
Biscay-North Pyrenean fault, 107
Bituminous coal, **102**,103
Black Sea, 39,65
Black shale, 40,168,177,179
Blake Plateau, 16
Blanket peat, 100
Block-fault, 55,78,86,87,177
Blueschist metamorphism, 10
Boca de Vitrila, **162**,164
Boghead coals, 102
Bohemia, 105,**106**
Bolivia, 140,**141**,153
Bornite, 139
Boron, 172,180
Brazil, 15,104,110,**156**,197
Brazilian Shield, 61
Brent Field, **56**,88
Brines, 38
   ores of ocean crust, 127,128,**131**
   sabkha, 196
   sedimentary ores, 157,169,170,171
British Columbia, 152,181
British Isles, 39-40,111,112,**175**
Brittany, 105,106
Burial, of sediments, 42,44,48,93,115
   depth of, 42,**43**
Burma, 60,61,**62**,140,141

207

Index

Calc-alkaline rocks, 11,136
Calderas, 140
Caledonian faults, 174
Caledonian Mountains, 112,137
Caledonide margins, 171
California, **63**,64,107,152
California, Gulf of, 4,16,129
Camarouns, 180
Cambrian Period, 107,181
Canada, 3,169,173,197
   Arctic, 16,163
   coal, 101
   ore deposits, 158,169,**170**,173,180
   petroleum basins, 64,77
   Western basin, 51
Canadian Arctic Basin, 15
Canadian Shield, 113
Cannel, 102
Cantabrian coalfield, **106**
Cantabrian Mountains, 107
Cape Basins, 197
Capillary pressure, 90
Cap-rock, 33,55,*90-91*
Carajas, **155**
Carbon cycle, 34,**35**
Carbon fixation, 42
Carbonates, 74,77,78,80,*198-199*
   reservoirs, 76-80
Carbonatites, 202,203
Carboniferous Age, 13,**14**,55,79
   coal, 55,79,99,103-107,**108**,109,111-115
   ores, 169,173,174,**175**,176
Caribbean Sea, 137
Caspian Sea, 201
Cassiterite, 151
Catagenesis, 42,45,46
Cenozoic, 60
Central African Copperbelt, 161
Central Grabens, 163
Cerro Bolivar Mine, **155**
Chalcopyrite, 129,131,139
Channel meandering, 74
Charcoal, 102,110
Chert, 175,177,181
Chile, 140
China, 13,14,91
Chloride complexes, metallic, 169,172,196, 200
Chromite ores, 119,120,121,*123-127*
Cincinnati Arch, 113
Clarain, 101,103
Climates, 38,39,41,78,90
   coal formation, 100,104,109,114
Climatic belt, 194

Coal, 14,*99-115*,**196**,201
   bituminous, **102**,103
   boghead, 102
   depositional environment, 104-114
   formation, 114-115
   humic, 101
   process of coalification, 101-104
   rank of, 103
Coal-gas, 103
Coastal Ranges, 58
Cobalt, 129,133,161,166
Cocos Ridges, 4
Coeur d'Alene, 180
Coke, 103
Collison(s), 138,144,201
   continental, 115,166
Collision belts, 137,138,*143-144*
Collision orogeny, 3,115,177
Colorado, 143
Colorado Plateau, **160**,166
Compaction, **43**,47,69
Conduction, 21
Continent collision, 115,166
Continental crust, 11,12,13,112
Continental margins, 54,115
Continental separation, 12, *13-16*
Convection, 2,21,23
Cook Inlet, Alaska, 59,**60**,84,**85**
Copper
   ocean crust, deposits, 119,127,129,130, 131,133
   Red-bed, **160**,*160-166*,167,171,172,179
   sedimentary, deposits, 159,166,168,172, 174-176,178-181,184,185
   subduction, association, 137,140,143,144
Copper Harbour Conglomerate, 163
Copper sulphide, 119,121,127,139,163, 179,201
Coppermine River, 162,163
Coral, 76,197
Coral reefs, 41,76,**196**,197
Cordillera, Western, 137,**141**
Core, 19,21
Coronation geosynclinal evolution, 12
Covellite, 139
Cratons, 22
Cretaceous Period, 12,15,**16**,95,194,197, 198,200
   coal, 99
   ore deposits, 136,143,172,173,183
   petroleum
     reservoirs, 55,60,**61**,**65**,66,75
     sources, 39-40,41,42
     traps, 83

Crust, 22
   continental, 11,12-13,112
   destruction, *see* Subduction
   oceanic, 1,7,*119-133*
   spreading, *52-***56**,**68**
Cuprite, 139
Cuticle, 102
Cyclothem, 101,111,114
Cyprus, **130**,**131**
Czechoslovakia, 105

Dakota, 109
Damara, 144
Dandeldhura granite, 143
Deep Sea Drilling Project, 41,92
Deep-sea environment, 91-94
Deep-sea fans, 74,**75**
Deep-water shale, 68,180
Deltas, 200
   coal, 105,111
   ore deposits, 152,180
   petroleum, 68,**69**,74,**75**,78,79,87,94
Desert, 199
Devonian Period, 77,99,107,114,169,**170**, 181,182,200
Diagenesis, 42,157,159
Diamonds, 12,151,203
Diapirs, 120,195,200,202
Diorites, 136,144
Dismal swamp, USA, 114
Dolomite, 77,198
Dolomitisation, 171
Donetz coal basin, **108**,114
Dunes, 74,**75**,199
Dunite, 120,121,123,**124**,**126**
Dunkard Group, 113
Durain, 101,103
Durham, 104

Earth, rotational axis, inclination, 193,194
Earthquakes, 2,**3**,5,9,204
Earth-sun distance, 193
East African Rift, 53,203
East Pacific Rise, 4,121,129
Ecuador, 60
Eire, 175
Ekofisk, 55,**56**
El Salvador porphyry deposit, 164
Eldfisk, **56**
Elk Point, **170**
England, 112,162
Enstatite, 124,125

Index

Eocambrian glaciations, 193,194
Eocene, 15,16
Equatorial deposits, *195*
Equatorial rain belt, 194,195
Euler poles, **5**,8,15
Euphotic zone, 35,**36**,37,38
Europe, 13,15,16,197,198
    coal, 105,**106**,113,115
    sedimentary ores, 162,166,167,176
Eustatic effect, 93
Evaporites, 128,194,**196**
    petroleum, 54,55,**66**,67,78,87,88,91
    sedimentary ores, 161,**162**,163,165,166, 168-171,173,174,177,178,185,186
Evaporitic deposits, *195-197*
Everglades swamp, 114

Failed arm, *see* Aulacogen
Faults, 47,58,78,92,107,115,120,173,178, 179,184,204
    block, 55,78,86,87,177
    transform, *see* Transform faults
    traps, 82,87-90
Fiji, 137
Flood plain, 105
Florida, 152
Fluid inclusions, 131,175
Fluid pressure, 46
Fluorine, 172,173,174
Fluorite, 143,168
Flysch, 137,143,173
Folding, 115
Fore-arc basins, *57-58*,88
Fore-deep basins, 89
Forties Field, 55,**56**
France, 105,106,107,144
Fundy Basin, Nova Scotia, 107
Fungi, 100
Fusinite, 102,110

Gabbro, 120,121,**122**,125
Gabon, 54,66
Gabon Basin, **66**
Galapagos Islands, 4,129
Galena, 179,181
Gas, *31-96*
Gays River, Nova Scotia, 176
Geosyncline, **155**
Geothermal gradient, 43,44,48,63,64
Geothermal heat, 34,39,43
Germany, 55,111,112,176
Ghawar oilfield, 48

Gippsland, 100
Glacial deposits, 200-201
Glaciation, **10**,14,110,111,158
    Eocambrian, 193,194
Glaciers, 199
Glass, 200
Gomar Challenger, 92
Glossopteridae, 109
Glossopteris, 110
Gold, 131,140,151,152,**153**,154,155
Gondwana, **14**
Gondwana ice sheet, 13
Gondwanaland, 13,15,201
    coalfields, **108**,109-111,114
Gortdrum, 174
Gossans,149
Granites, **136**,141,143,144,145,150,154
Granulite, 11,22
Gravel, 200
Great Slave Lake area, 166
Green River, Wyoming, 199
Greenland, 15,16,155
Greenschist, 131
Greenstones, 11,150,154
Grenvillean orogeny, 12
Greywackes, 150
Gypsum, 140,181

Haad Som Pan, 141
Halocline, 39,40
Halokinetic movements, 86
Hammersley Basin, **155**,156
Harzburgite, 120,121,**122**,**124**,125,**126**
Hawaii, 133
Heat
    geothermal, 34,39,43
    radiogenic, 20
Heat-flow, **2**,**3**,4,6,7,12,19,138,*202-203*,204
    coal, changes, 104
    ocean crust, 7,121,128,129
    petroleum generation, 33,34,44,55,61,93
    sedimentary ores, 150,174,176,180,181, 184
Haematite, 165,175
Hercynian Range, 104,105,106,143,162, 177,182
Hilt's Law, 103
Himalayas, 3,69,143
Hoggar Massif, 203
Honshu Island, 140
Hot spot, 110
Howard's Pass, 177,181,182
Humic coal, 101

Huronian, 12,154
HYC deposits, **178**,179
Hydrocarbons, *31-96*,115,201,202,204,205
    carbonates and, 198
    evaporitic deposits, 195,196
    ratios, extractable, 44
    in sandstone, 199
    sedimentary ore deposits, 165,168,174, 175,182,184
    see also Petroleum
Hydrothermal circulation systems, **2**
Hydrothermal fluids, 7
Hydrothermal ores, *127-132*
Hydrothermal processes, 121,132

Iberian Pyrite belt, 182
Ice age, 200
Iceland, 6
Idaho, 199
Ignimbrites, 140
Illinois coalfield, 113
Indefatigable Field, **56**
India, 13,15,101,111,183
Indian Ocean, 18,123
Inter-arc basins, *59*,88
Interior basins, 115
Intermontane basins, 88,105
Interstellar dust clouds, 193
Intraplate basins, 52,*64-70*
Iran, 57,66,**67**,69,77,**84**
Iraq, **67**,76,77
Irish zinc-lead deposits, 168,172,*174-177*
Iron, 121,159,180,184
Iron formations, *154-159*,175,185
Iron oxide, 122,127,129,131
Iron sulphide, 127,139,163,179,183
Island-arc, 2,**3**,10,136
Isostacy, 7
Isostatic adjustments, 21,23,204
Isua, 155

Jacobina, 154
Jan Mayen Ridge, 16
Japan, 10,137,140,143
Juan da Fuca Ridge, 4
Jurassic Period, 14,15,55,66,80,95,193, 194,199
    coal, 99

Kaolin, 195
Karroo basalts, 11,14

Index

Karroo group coal deposits, 109,110,114
Katanga, 161
Katharina Marine band, 111
Keewanawan rift, 163,166
Kennecott, Alaska, 201
Kerogens, 42,46,47
Kimberlites, 12,202
Kimmeridge clay, 39
Kinta Valley, Malaya, 152
Klondike, 152
Kolyma Block, 16
Krivoy Rog, **155**
Kupferschiefer, 160,**162**,164
Kuroko-type deposits, 137,*139-140*
Kuwait, **67**,76
Kyushu Island, 143

Labrador geosyncline, **155**
Labrador Sea, 15,198
Laissvall, Sweden, 171
Lake Superior, **156**
Lakes, limnic, 100,104
Laterites, 149,195
Laurasia, 109,112,113,114
Laurentia, 13,14,15
Lead, 133
 -rich fluids, 175
 sedimentary deposits, 159,**160**,172,173, 179,180,183,185
  Irish zinc-lead deposits, *174-177*
  red-bed copper, 161,162,163,165
  shale-hosted, *177-184*
  zinc-lead deposits, *167-184*
 subduction, associated with, 140,143
Lead sulphide, 139,183,198
Libya, 83
Lignite, 103,109
Limestone, 76
Limnic coal basins, 100, 104,*104-111*
Lithosphere, **2**,**3**,17,18,**19**,92,203
 continental, 9-13,18,92,202
 oceanic, 4-9,119,**120**,123,**126**,135,203
 Precambrian, 20,21,22,23
Loading of sediment, 93
Lomonosov Ridge, 16
Lorraine, France, 105
Los Angeles Basin, 48,63,87
Low-velocity zone, seismic, 13,18,19
Lublin Basin, **108**
Ludworth Dyke, 103

McArthur Basin, 178
McArthur River, 177,**178**,181
MacDonald Fault, 170
Madagascar, 13,14
Magmas, 7,9-10,119,**120**,121,125,**126**,127, 138,173
 alkaline, 176
Magmatic arc, **136**,*136*
Magnetic anomalies, 5,7,8
Magnetisation, inclination, **196**
Magnetite, 140,151
Malaysia, 140,144,152
Mammoth coalfield, 113
Manganese aureole, 175
Manganese nodules, 119,*132-133*
Manganese oxide, 122,127,129
Mantle, 8,12,13,17,22,124,125,202,204
 low-velocity layer, seismic, 13,19
Maracaibo Basin, 65
Margins, continents, 54,115
Marquette, **156**
Mars, 21
Massachusetts, 107
Massif, Central, 105,**106**
Maturation, petroleum source rock, 42,48
Meggen deposits, 177,182
Mendips, 176
Mercaptans, 169
Mercury, 174,176
Mesozoic Period, 37,38,39,41
Metallic ores, 10
 see also Ores
Metallogenesis, 7
Metamorphic belts, 'paired', 10
Metamorphism, 103,110,115,150
Meteoric waters, 138
Methane, 45,46,103
Mexico, 137
Mexico, Gulf of, 13,14
Michigan, 163,199
Michigan Basin, **108**,113,204
Micrinite, 102
Microfractures, 46
Mid-Atlantic Ridge, 7,129
Mid-Continent coalfield, 113
Mid-European Sea, 108
Mid-ocean ridge, 119,**126**,132
Middle East Basin, 66,**67**,**68**,76,77,94,*95-96*,198,199
Middle Ground Shoal, 84,**85**
Miocene, 16
Mississippi Delta, 69,79,114
Mississippi Valley, 168,173,176,196
Molasse, 60,69,70,**141**

Molasse Basin, 69
Molybdenum, 133,137,143
Monsoons, 194,195
Montagne Noire, **106**
Montana, 109
Montmorillonite-type clay material, 45,46
Moon, 21
Moravian Silisian Basin, **106**
Moscow Basin, 64,**108**,114
Mount Isa, 177,179

Namibia (South West Africa), 109,144
Nares Straits, 15
Narragansett Basin, 107
Navan, 174
Nepal, 143
New Caledonia, 112
New England, 107,108
New Mexico, 166
New Red Sandstone, 167
New Zealand, 13,15,152,**153**
Newfoundland, 132
Nickel, 129,133,166
Nickel sulphide, 127
Niger Delta, 69,75,**76**,79,87,114
Nonesuch Shale, 163
Nontronite, 127,129,131
North America, 13,14,15,16,136,138,180
 coal, 109,113
North Sea
 coal basins, 103,111,112,115
 ore deposits, 162,163
 petroleum, 37,54,55,**56**,76,80,87,103
Norwegian-Greenland sea, 16
Nova Scotia, 107,**108**,176

Ocean crust, 1,7
 ore deposits of, *119-133*
 seismic surveys of, 7
Ocean-floor spreading, 5-6,11,23,119,**120**, 123
Ocean stratification, 38,39,40,41
Oceanic ridge, *see* Ridges, oceanic
Oceans, deep, 38
Oceans, vertical mixing in, **36**,37,38,39,40
Ochres, 131
Oil, *31-96*,173,**196**
Oil generation threshold, 93
Oil shale, 167
Old Red Sandstone, 167,174,176
Oligocene Period, 15
Olivine, 123,124,125,126,127

Index

Olomoro Field, **76**
Oman, 77, 132
Open-shelf environments, 40
Ophiolite complexes, 119,120,121,123,125, 127
  sulphides in, **130**-132
Ordovician, 169,181,183
Ores
  deposits
    of ocean crust, *119-133*
    sedimentary, *149-186*
    subduction and, *135-145*
  metallic, 10
  volcanogenic, 4
Oriente Basin, **61**
Orogenic activity, 2
  collisional, *see* Collision
Orogenic belts, 11
Oslo Graben, 163
Ouachita Basin, 173
Outer-arc magmatic belts, *137,143*
Overpressures, 43,45,46,69
Oweh Field, **76**
Ozark, 173

Pacific ocean, 14,48,123,133
  Western basins, 4
Pacific ridge, 7
Pakistan, 140
Palaeo-Eocene, 16
Palaeoclimate, 23,*193-201*,205
  coal, 108,109,114-115
  petroleum, 78,79,91
  sedimentary ore deposit, *149-186*
Palaeogeography, 4,33,95,108,150,193, 194,200,201,*205*
Palaeolatitude, 78,79,91,114-115,166,*193-201*
Palaeomagnetism, 23,205
Palaeozoic Period, 14,43,44,64,95
Pangaea, 14
Pannonian Basin, 65
Papua-New Guinea, 137
Paradox Basin, Utah, 199
Paraffins, 34
Paralic environment, 104,**106**,*111-114*
Paraná Basin, Brazil, 110
Partial melt, zone of, **120**
Peace River Arch, 170
Peat, 99,100,101,103,104,105,110,111, 114
Pennine Range, 176
Pennsylvania, USA, 104,111,113

Peridotite, 120
Permeability, 74
Permian age, 13,**14**,54,57,201
  coal, 105,106,109,110,113,114
  ore deposits, 162,166
  petroleum, 54,55,57,76,77,95
Permo-Triassic Period, 32
Persian Gulf, **67**,**68**
Peru, **59**,60,**61**
Peru-Chile Trench, 203
Petrofabric, 123,124,125
Petroleum, *31-96*,169,**170**,173
  cracking of, *43*
  deep-sea environment, 91-94
  formation, *42-44*
  migration, *44-48*
    primary, 44-47
    secondary, 47-48
  nature of, 33-34
  reservoirs, 33,43,47,*51-70*,*73-90*
  sources, *33*,*34-42*,48
    basinal model, 40-42
  traps, 33,47,*80-91*
Phanerozoic Period, 11,150
Phase changes, 17,21
Philippines, 132,137
Photosynthesis, 34,**35**,99,158
Phytoplankton, 35,40
Pillow lava, 131
Pine Point, **170**
Piper Field, **56**
Pittsburgh, USA, 111
Placers, deposits, 141,144,*151-154*,205
Plants, 99
Plaster, 196
Plate margins, *1-13*
  constructive, 2,4-9
  destructive, 2,9-13
Plate tectonics, *1-24*,51
  basic theory of, 1-4
  coal formation, 114-115
  mechanism of, **16**,*16-20*
  Precambrian, *20-24*
Platinum, 127,151
Po Valley, 69
Poland, 105,111,112,162
Pollen, 100,110
Polymetallic deposits, 159
Pore space, 47
Pore waters, 45,159
Porosity, 149,171
  effective, **73**,74
  secondary, 77
Porphyry deposits, *137-139*,144

Potash fertiliser, 196
Potassium, 18
Precambrian Period, 20
Precambrian Shield, 104
Productivity, primary, 35,36,37
Proterozoic Period, 11,21,22,23,150,185
  iron formations, 155,157,158,159
  placer deposits, **153**,154,185
  red-bed copper, 160,161,163
  uranium-vanadium-copper, 167
  zinc-lead deposit, 172,177,178,180
Prudhoe Bay, **32**,**83**
Purcell sequence, 180
Pycnocline, **36**,38,39,**40**,41
Pyrenees, 177,183
Pyrite, 129,139,161,163,164,174,179,181
Pyroxenite, 120
Pyrrhotite, 129

Qatar, **67**
Quadrilatero Ferrifero, **155**,**156**
Quaternary ice age, 200

Radstock coalfield, 113
Rain belt, equatorial, 194,195
Rain forests, equatorial, **14**
Rain-shadow areas, 194
Rajmahal Traps, 14
Rapitan Iron Formation, **155**,158
Rare earth elements, 203
Recent age
  basins, 63
  coalification, 103
Red-bed copper, **160**,*160-166*,167,171,172, 179
Red beds, 194,199
Red Sandstone, 167
Red Sea, 16,38,53,54,55,92,128
  Atlantis Deep, 127,128
Reefs
  biohermal, *197*
  coral, 41,76,**196**,197
  petroleum, 40,55,76,77,80,86,91
  sedimentary ores, 170,171,179
Regressions, 66,67,**68**,74,79
Rengiu oilfield, 91
Reocin deposits, 183
Resin, 102
Resistate minerals, 151
Rhine Graben, 203
Rhyolite, 140
Ridge-crest faults, 129

Index

Ridges, oceanic, 1,**2**,*4-9*,17,41,202
   heat-flow, 6,202
   mid-oceanic, 119,**126**,132
Rift, 2,52,**53**,54,115,185,186,201,205
   petroleum, 78,79,88,201
   sedimentary ores, 161,162,166,181,185,186
Rift valley, 5,6,17,53,203
Rockall Bank, 16
Rocky Mountains, 109,**170**
Roll-over anticlines, 69,87
Rossing deposit, Namibia, 144
Rotation, Euler pole, **5**,8
Rotliegendes Formation, 55,76
Russian Platform, 114
Rutile, 151

Saar coal basin, 105,**106**
Sabkha, 67,195,196,197
Sahara, 203
Salt domes, **86**
Salt plugs, **86**
San Andreas Fault, **63**,107
Sands, 78,79,200
Sandstone, **73**,74,77,*199-200*
   red, 167,174,176
   reservoirs, 74-76
   uranium-vanadium-copper deposit, *166-167*
Sapropel, 100,101,102
Sardinia, 182
Sargasso Sea, 36
Sargipali, India, 177,183
Saskatchewan, 109,167,169
Saudi Arabia, 66,**67**,**68**,76,77,86,128
Scandinavia, 171
Scotia Sea, 4
Scotland, 112,137
Sea floor, 181
Sea-floor spreading, 5-6,12-13,23,119,**120**,123
Sea-levels of Cretaceous, 41
Sea-mounts, 80
Seal, petroleum, 47,*90-91*
Sedimentary basins, 33,43,47,*51-70*
   classification of, 51-**52**
   crustal destruction, **52**,*57-61*
   crustal spreading, *52-***56**,**68**
   intraplate, 52,*64-70*
   transform faulting, 52,*61-64*,78,89
Sedimentary ores, *149-186*
   primary, 151-159
   secondary, 159-184

Seismic activity, 7,9
Seismic reflection profiling, 91
Serpentine, 120,123,124
Shales, 200
   black, 40,168,177,179
   deep-water, 68,180
   oil, 167
Shearing, 120
Shimanto Belt, Japan, 137
Siberia, 13,14,15,16,114,152
   East, 104
Siberian Traps, Triassic, 11
Silled basin, 197
Silurian Period, 99,153,181,182,183
Silver, 140,174,180,181
Silvermine, 174
Siwalik Trough, 69
Solomon Islands, 137
Source rocks, petroleum, 33,34,48,95
South Africa, 109,110,167
South America, 13,110,**141**,**142**
South West Africa, 109,144
Spain, 91,**106**,107,112,177
Sphalerite, 129,131,174,179,181
Spore coloration, 42,44
Spores, 100,102,110
Stratiform ores, 161
Stromatolite, 172
Strontium, 172,203
Strontium ratios, 138,144,**145**
Subduction, 18,23,**52**,*57-61*,105,138,143,203
   ore deposits associated, *135-145*
Subduction zones, 2,**3**,*9-13*,105,112,202,203,205
Subsidence, 101
Sudan, 128
Suez, Gulf of, 53,54
Sullivan deposits, 177,180,181
Sulphur, 159,161,164-166,168,169,173,176,185,196
Sumatra, 59, 61,**62**,137,144
Supergene zones, 139
Superphosphates, 196
Surface temperature, 20-21
Swamp-forests, 99
Swamps, 101,111,112,114
Sydney Basin, 114
Syngenetic deposits, 179

Taiwan, 137
Tasman Sea, 15
Tasmanian basalts, 14

Temperature
   petroleum source rock, 42,43
   surface, 20-21
Tertiary Period, 58,63,64,**85**,103
Tethys Ocean, 95,201
Texas, 197
Thailand, 137,140,141,144
Thermocline, 38,39
Thorium, 18,167
Tin, 140,141,143,144,152,153,180
Titanium, 133,152
Tonalitic inclusions, 136
Topography, **2**,6
Torbanite, 102
Trais, 113
Transcurrent motion, 107
Transform faults, 5,8,**52**,*61-64*,78,89
Transgressions, 41,66,67,**68**,74,78,101,161,163
Transvaal, 109,110,**155**,**156**
Traps, petroleum, 33,47,*80-91*
Trench, 1,**3**
Triassic Period, 95,99,105,109,144
Triassic Siberian Traps, 11
Triple junctions 8,**9**
Tristan da Cunha, 6
Troodos ophiolite, Cyprus, 130,132
Tungsten, 141,143
Turbidites, 58,78
Turonian, 15
Tynagh, 174

Ukraine, 114
Umber, 129
Unconformities in traps, 83
Under-compaction, 69
United Arab Emirates, **67**
United States of America, 4,64
Upper Radnice, 105
Upwelling, **37**
Ural Mountains, 14,114,137
Uranium, 18,143,144,154,**160**,*166-167*,199
Utah, 166,199

Vanadium, **160**,*166-167*,199
Vanderbeckei, 11
Variscan Mountains, 14
Vassbo, Sweden, 171
Venus, 21
Verkhoyansk Mountains, 15
Vienna Basin, 65
Viking Graben, 163

Index

Vitrinite, 101,102,103
  reflectance, 42,44
Volcanic arc, **57**
Volcanogenic ores, 4
Vosges, 105,**106**

Wales, South, 104,112,113
Walvis-Rio Grande Rise, 197
Water saturation, 73
Weardale granite, 104
West Sole, **56**
Westphalia, **108**
White Pine Copper District, 160,**162**,163, 164
Williston Basin, USA, 64,**65**,86,89
Witwatersrand, **153**,154
Wyoming, 109,166,199

Yoked basins, 105
Yukon, 152,181
Yukon-Northwest Territories Border, 181

Zagros Basin, 57
Zagros Mountains, **67**,**68**,70,77,**84**,95
Zaire copperbelt, 161
Zambia, 161,**162**,164
Zambian Copperbelt, 160,161
Zechstein Sea, 162,163
Zinc, 127,129,140,143,159,172,173,179, 180,182,185
  red-bed copper, **160**,161,162,163,165
Zinc-lead deposits, *167-184*
  in Devonian Canada, **170**
  Irish type, *174-177*
  in shale, 177-184
Zinc sulphide, 199,121,139,183,198
Zooplankton, 34